Floating Offshore Energy Devices
GREENER

Floating Offshore Energy Devices GREENER 2019
Trinity College Dublin, Dublin, Ireland; 24-25 Sep, 2019

Editors
Ciarán Mc Goldrick, Meriel Huggard and Biswajit Basu

Trinity College Dublin, Dublin, Ireland
https://www.tcd.ie

Peer review statement

All papers published in this volume of "Materials Research Proceedings" have been peer reviewed. The process of peer review was initiated and overseen by the above proceedings editors. All reviews were conducted by expert referees in accordance to Materials Research Forum LLC high standards.

Published under License by **Materials Research Forum LLC**
Millersville, PA 17551, USA

Published as part of the proceedings series
Materials Research Proceedings
Volume 20 (2022)

ISSN 2474-3941 (Print)
ISSN 2474-395X (Online)

ISBN 978-1-64490-172-4 (Print)
ISBN 978-1-64490-173-1 (eBook)

This book contains information obtained from authentic and highly regarded sources. Reasonable efforts have been made to publish reliable data and information, but the author and publisher cannot assume responsibility for the validity of all materials or the consequences of their use. The authors and publishers have attempted to trace the copyright holders of all material reproduced in this publication and apologize to copyright holders if permission to publish in this form has not been obtained. If any copyright material has not been acknowledged please write and let us know so we may rectify in any future reprint.

Distributed worldwide by

Materials Research Forum LLC
105 Springdale Lane
Millersville, PA 17551
USA
http://www.mrforum.com

Manufactured in the United States of America
10 9 8 7 6 5 4 3 2 1

Table of Contents

Preface

The first GREENER conference was held in Trinity College Dublin in September 2019.

GREENER is a biennial initiative of the **ICONN** project, and draws together a select group of leading scientists, innovators, industrialists and decision makers from all over the world to discuss and share state-of-the-art research, innovation and practice in the development and deployment of **present and future offshore energy machines and devices.**

The ICONN project was a European Commission supported Industrial Research Training initiative (Grant Number: 675659) coordinated by Trinity College Dublin, Ireland. Under the auspices of the Marie Skłodowska-Curie Innovative Training Networks programme the ICONN consortium helped train, develop and build future European Research and Innovation capacity – through the fostering and development of both human capital and cutting-edge wind and wave energy research and training innovations. The GREENER conference is one of the enduring communication activities of the ICONN project.

This genesis of the GREENER conference is evident in the papers gathered in this volume. The papers span a diverse range of advanced technical topics relating to offshore energy, whilst simultaneously embracing the essential excellence in training underpinning of the MSCA-EID ICONN project.

The early papers in the volume introduce both offshore wind and wave devices, and provide valuable overviews of the state-of-the-art in the respective domains.

Thereafter the accepted papers document specific advances and innovations in the modelling, design, control, operation and testing of offshore energy machines – having particular regard for the effect and impact of their marine environment on both the devices and their generating efficiencies.

The Editors would like to express their sincere thanks to the following for their assistance and support in the organization and running of the 2019 Greener conference.

Sponsors

Trinity College Dublin

Trinity
College
Dublin

The University of Dublin

European Commission MSCA-ITN Programme. European Industrial Doctorate Grant Number: 675659

Committee

General Chairs
Ciarán Mc Goldrick, Biswajit Basu

Organising Committee
Meriel Huggard, Breiffni Fitzgerald, Ciarán Mc Goldrick, Biswajit Basu, Giacomo Politi, Tao Sun (Trinity College Dublin)

Scientific Committee
Adrian Constantin, University of Vienna, Austria; Antonio Falcao, Instituto Superior Técnico, Portugal; Lars Andersen, Aarhus University, Denmark; Zili Zhang Aarhus University, Denmark;
Jan Høsberg, Technical University of Denmark, Denmark; Wieslaw Staszewski, AGH University of Science and Technology, Poland; Jianbing Chen, Tongji University, China; J Lie, Tongji University, China; Amélie Têtu, Aalborg Universitet, Denmark; Srinivas Guntur, Siemens USA; Matthew Lackner, University of Massachusetts, USA; Andrea Staino, Alstom, Paris, France;
John Ringwood, Maynooth University, Ireland; Ricardo Carbajo, CeADAR, UCD, Ireland

ICONN Consortium
Søren R. K. Nielsen (Aalborg Universitet), Jens Peter Kofoed (Aalborg Universitet), Sarah Thomas (Floating Power Plant), Anders Køhler (Floating Power Plant), Chris McConville (Floating Power Plant), Ciaran Mc Goldrick (Trinity College Dublin), Biswajit Basu (Trinity College Dublin), Meriel Huggard (Trinity College Dublin), Breiffni Fitzgerald (Trinity College Dublin)

Floating Offshore Energy Devices   Materials Research Forum LLC
Materials Research Proceedings **20** (2022) 1-9   https://doi.org/10.21741/9781644901731-1

Overview on Oscillating Water Column Devices

António F.O. Falcão

IDMEC/LAETA, Instituto Superior Técnico, Universidade de Lisboa, 1049-001 Lisbon, Portugal

antonio.falcao@tecnico.ulisboa.pt

Keywords: Wave Energy, Oscillating Water Column, Air Turbines, Control

Abstract. Oscillating-water-column (OWC) converters, of fixed structure or floating, are an important class of wave energy devices. A large part of wave energy converter prototypes deployed so far into the sea are of OWC type. The paper presents a review of recent advances in OWC technology, including sea-tested prototypes and plants, new concepts, air turbines, model testing techniques and control.

Introduction

The sea waves are a vast, practically untapped, renewable energy resource. Many concepts and technologies for their utilization have been proposed and developed with varying success. They have been classified with respect to their basic concept and to their location with respect to the shoreline. The oscillating water column (OWC) is widely regarded as the simplest and most frequently adopted type of wave energy converter. More OWC prototypes have been deployed and tested in the real sea than of any other type of wave energy device. The OWC converters may be bottom standing, integrated into a breakwater or floating. They consist of a hollow structure, fixed or floating, open at its submerged part, within which the air trapped above the inner free-surface is alternately compressed and decompressed by wave action. In almost all cases, the air chamber is connected to the atmosphere by a self-rectifying turbine. An extensive review of OWCs can be found in [1].

Resonance plays a central role in almost all wave energy converter concepts if a satisfactory efficiency is to be attained. This involves one or more oscillating bodies or oscillating masses of water (water columns). A single-oscillating-body converter reacts against a fixed frame of reference, in general the sea bottom or a bottom standing structure. This may be avoided in floating devices consisting of two or more bodies that are mechanically inter-connected (hinge or other connection). In OWC devices, the water column acts as an oscillating body without the need of any mechanical connection.

In almost every case, the power take-off system (PTO) of an OWC converter is relatively conventional and reliable: an air turbine (in most cases a self-rectifying version) directly driving an electrical generator, located above sea water level. A dielectric elastomeric membrane generator capable of converting deformation into electrical energy has recently been proposed as an alternative to the air-turbine-generator set [2].

Recent sea-tested OWC prototypes and plants

OWC prototypes have been deployed into the sea since the 1970s, an early case being the Kaimei floating vessel in Japan. Here, only recent realizations are mentioned.

A bottom-standing plant was deployed in 2016 near the coast of Jeju island, South Korea, Fig. 1. It is equipped with two self-rectifying axial-flow impulse turbines of 250 kW rated power each.

Floating Offshore Energy Devices
Materials Research Proceedings 20 (2022) 1-9

Materials Research Forum LLC
https://doi.org/10.21741/9781644901731-1

Fig. 1. Bottom-standing OWC in Jeju island, South Korea, completed in 2016. It is equipped with two 250 kW self-rectifying axial-flow impulse air turbines.

The integration of wave energy converters into harbour protection structures has been considered an interesting option since the early times of wave energy development. The costs of the dual-purpose structure are shared, and the access for construction, installation and maintenance are made easier. OWCs are especially appropriate for integration into breakwaters.

A breakwater with 16 OWCs was constructed at Mutriku harbour, in Basque Country, Spain. The plant, completed about 2012, was equipped with 16 bi-plane Wells turbine-generator sets rated 18.5 kW each (Fig. 2).

Fig. 2. Mutriku harbour breakwater with 16 OWCs, completed about 2012.

A much longer breakwater was constructed at the port of Civitavecchia, Italy, integrating 124 OWCs (completed in 2016) (Fig. 3). Only one turbine (bi-planeWells type) was (temporarily) installed. The U-shape of the Civitavecchia OWCs, invented by P. Boccotti [3], allows a longer (and more easily resonant) OWC, while keeping the mouth close to the sea surface, Fig. 4. The U-OWC breakwater concept is planned to be replicated elsewhere in Italy.

Several floating OWC concepts have been proposed and studied so far. Here we mention two that reached the stage of sea tested prototype in the last decade or so.

The backward-bent-ducted-buoy was proposed in the mid-1980s by Yoshio Masuda. A large model, at scale about 1:4th, was tested in Galway Bay, Ireland. It was equipped firstly with a Wells turbine and later with an axial-flow impulse turbine (Fig. 5). A full-scaled prototype was very recently constructed in Portland, Oregon, USA, to be deployed in Hawaii (Fig. 5).

Materials Research Forum LLC
https://doi.org/10.21741/9781644901731-1

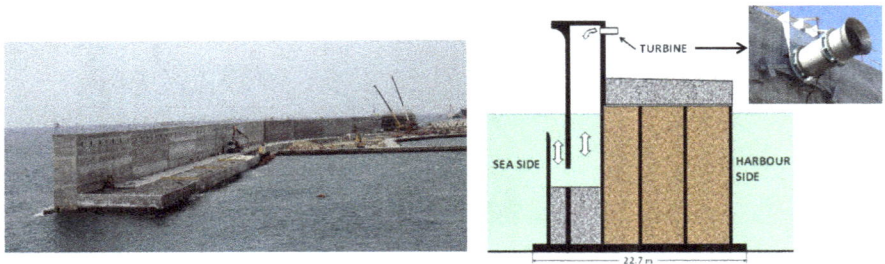

Fig. 3. Breakwater at Civitavecchia with 124 OWCs (2016).

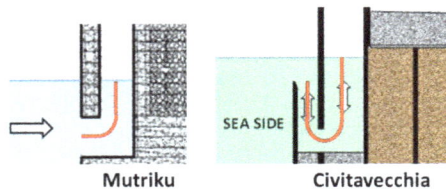

Fig 4. Cross-section comparison of the Mutriku breakwater "conventional" OWC and the U-shaped OWC of Civitavecchia breakwater.

Fig. 5. BBDB prototypes. Left: 1:4th scale being tested in Galway Bay, Ireland, about 2008. Right: full-sized prototype in shipyard, in Portland, Oregon (2019), for deployment in Hawaii.

Another floating OWC concept is the spar-buoy OWC converter, consisting of an axisymmetric floater with a long coaxial vertical tube, open to the sea at its bottom end, within which is located the water column. This concept was extensively studied theoretically, numerically and in wave tank in the last few years. A prototype, scaled about 1:3rd, was built and tested at the BiMAP test site, Basque Country, Spain, in 2018-2019. The converter was equipped with a 30 kW bi-radial self-rectifying air turbine. The turbine-generator set had been previously tested for one year at one of the OWCs of the Mutriku breakwater.

Since the lower opening of the spar-buoy tube is deeply submerged (typically of the order of 30 m below the sea surface), the wave energy is essentially absorbed through the interaction

Materials Research Forum LLC
https://doi.org/10.21741/9781644901731-1

between of the oscillating floater and the surrounding waves. A concept in which the opposite situation occurs is the co-axial tube OWC. It may be regarded as a floating axisymmetric version of Boccotti's U-OWC (sea above), Fig. 7. Here, the facing-up opening of the tube is close to the sea surface. Besides, since the water plane area (i.e. the annular cross-sectional area of the inner tube wall at sea water surface level) is very small, the hydrostatic restoring force is also very small, and so is the frequency of the free oscillations of the floating structure that behaves as a semisubmersible structure. For this reason, the heave and pitch oscillations are very weakly excited by sea waves, which makes this kind of wave energy converter appropriate to fit multiuse floating platforms. A rigidly-connected array of five such OWCs was model-tested in 2017 within the framework of the H2020 project WETFEET (Fig. 7).

Fig. 6. Marmok-A-5 spar-buoy OWC tested at the BiMEP test site, Basque Country, Spain, 2019. The converter is equipped with a 30 kW bi-radial self-rectifying air turbine (see Fig. 8).

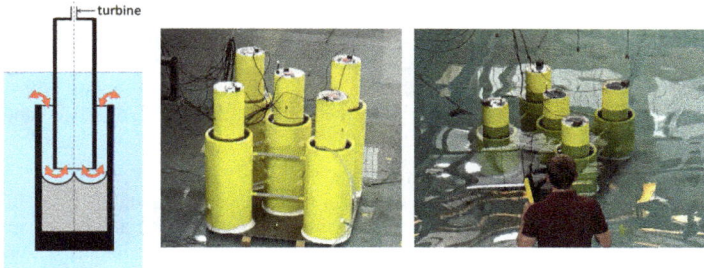

Fig. 7. Co-axial-tube OWC (left); five-OWC rigidly-connected array model (centre); and array being tested in the wave tank of the Coastal Laboratory, University of Plymouth (2017) (right).

Air turbines

An OWC converter is in general equipped with an air turbine coupled to a conventional electrical generator. This may be regarded as a simple and reliable type of power take-off system and is one of the attractive features of the OWC concept. If rectifying valves are to be avoided (which has been the case except in small wave-powered navigation buoys), the turbine must be self-rectifying, i.e., its rotational speed direction remains unchanged when the air flow is reversed by the reciprocating motion of the air column. Two types of such turbines were proposed and patented in the mid-1970s: the Wells turbine and the axial-flow impulse turbine, with variants of both. In spite of their limitations in terms of aerodynamic performance, they remain popular in OWC applications due to their mechanical simplicity and low cost. More sophisticated and efficient self-rectifying air turbines were developed in recent years.

Here, we mention especially the biradial turbine. This is an impulse turbine that is symmetrical with respect to a plane perpendicular to its axis of rotation. The flow into the rotor is radial centripetal and the flow out of the rotor is radial centrifugal. The rotor is surrounded by a pair of radial-flow guide-vane systems, each one connected to the corresponding rotor opening by a duct whose walls are flat discs (Fig. 8). A 30 kW biradial turbine was tested at Mutriku and then installed the Marmok-A-5 spar-buoy OWC in 2018-2019 at the BiMAP test site, Basque Country, Spain (Fig. 6). The turbine was equipped with a fast axially-sliding valve (opening or closing time 0.2 s) capable of achieving phase control by latching (see below).

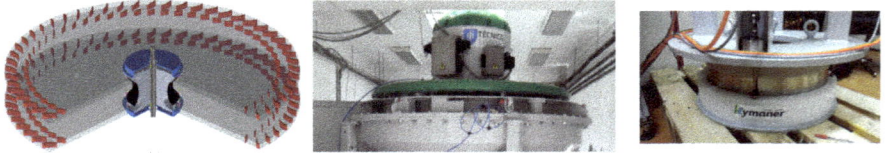

Fig. 8. Biradial air turbine. Perspective drawing showing the concept (left). Prototype before installion at the Marmok-A-5 spar-buoy OWC (centre), equipped with an axially-sliding high-speed valve (right).

The use of a conventional unidirectional air turbine requires a system of rectifying valves. This has been implemented in early small navigation buoys. Unidirectional turbines with rectifying valves were tested in Japan on the Kaimei floating vessel in 1978-80 and 1985-86, but the results were not encouraging [1]. The Tupperwave is a new concept of spar-buoy OWC equipped with a unidirectional turbine and check valves. The air flows in closed circuit, with low- and high-pressure reservoirs. Model tests were performed in wave tank with an orifice simulating the turbine [5]. The valves seem to remain a major problem.

The spring-like air compressibility effect
The volume of the air chamber of an OWC converter should be large enough to avoid ingestion of green water by the air turbine under rough sea conditions. Typical design values of the air chamber volume divided by the area of the OWC free surface range between 3 and 8m. The spring-like effect of air compressibility in the chamber is related to the pressure-density relationship, and increases with chamber volume. Such effect is important in a full-sized OWC converter. For such effect to be adequately simulated in model testing (Froude linear scale $\varepsilon < 1$ for the submerged parts of the converter), the ratio between the air chamber volume of model and prototype has to be equal to the square ε^2, not the cube ε^3, of the scale [6]. This implies a much larger air volume in the model which in general requires an additional rigid-walled air reservoir connected to the model air chamber. This rule is rarely implemented in model testing of OWCs, which means that most experimental results (published or unpublished) from OWC model testing could be affected by significant errors.

The compressible air in the chamber acts as a spring in series with the damping effect provided by the turbine. This produces a reactive effect that modifies (increases) the resonance frequency and consequently the frequency response of the converter. This effect may be unfavorable or (more rarely) favorable in terms of wave energy absorption (depending on incident wave frequency), but should not be ignored [6].

Floating Offshore Energy Devices
Materials Research Proceedings **20** (2022) 1-9

Materials Research Forum LLC
https://doi.org/10.21741/9781644901731-1

Control

Most wave energy converters, including especially OWCs, perform more efficiently near resonance conditions. Since real sea waves are irregular (rather than purely sinusoidal) and sea states vary widely along the year, control plays an essential role in converter performance. Generally control is implemented on the power take-off system (PTO), and so control strategies must be adapted to the converter type and especially the mechanical/electrical arrangement of the PTO. A wide range of control methods have been proposed and adopted [7,8].

In OWC converters, the PTO consists of an air turbine driving an electrical generator. The wave-to-wire efficiency of an OWC converter involves three processes: (i) hydrodynamics of wave energy absorption, (ii) aerodynamic performance of the turbine (this may include losses at non-return valves if the turbine is unidirectional), and (iii) performance of the electrical equipment. All three processes are coupled though the rotational speed of the turbine-generator set, and so control of the OWC converter relies largely on rotational speed control.

Phase control by latching was proposed in 1978 by Falnes and Budal to improve the wave energy absorption by oscillating body converters (especially point absorbers). It consists in latching the body in a fixed position during certain intervals of the oscillation cycle. Extending the latching control strategy to OWCs requires to air flow to be stopped during certain intervals of time; this requires a fast acting valve. Because in an OWC air is compressible, latching can be activated at any time without causing large peak forces, which makes it more versatile. For the same reasons why unidirectional air turbines have been unpopular (because valves are needed), also phase control by latching of OWCs has not been seriously considered until recently. The high-speed sliding valve that integrates the new biradial turbine (see above) may be used to successfully implement phase control by latching [9]. The numerical results in Fig. 9 show how latching control may dramatically increase the regular wave energy absorption by a spar-buoy OWC over a significant range of wave periods. Naturally, latching control may be effective only if the device's resonance frequency exceeds the wave frequency.

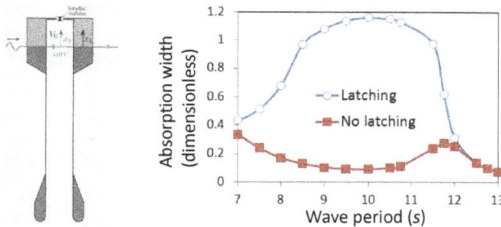

Fig. 9. Dimensionless absorption width of a spar-buoy OWC in regular waves with optimal latching control and without control. Buoy diameter 15 m, draft 38 m. Results from theoretical/numerical modelling [10].

Control of an OWC converter in most cases consists simply in controlling the rotational speed of the turbo-generator set through the electromagnetic torque L_e of the generator. This is mainly because the turbine aerodynamic efficiency depends strongly on its rotational speed. It should also be taken into account that kinetic energy is alternately stored in, and released from, the rotating masses (flywheel effect). An effective control algorithm is $L_e = a\Omega^{b-1}$, where L_e is the electromagnetic of the generator, Ω is rotational speed and the exponent b is (from turbomachinery non-dimensional analysis) approximately $b = 3$. Coefficient a depends on device configuration and size, and on turbine type and size, and must be optimized numerically or experimentally. This algorithm should be complemented to account for constraints related to maximum allowable rotational speed (especially if the turbine is of Wells type) and maximum allowable electrical power (especially in power electronics).

Electrical equipment control

Electrical generators are in general highly efficient machines (about 95%) in the power range above about 2/3 of the rated power. Below that level, the efficiency decreases markedly (see Fig. 10). The power absorbed from the waves varies greatly, depending on the sea state. The highly energetic sea states occur in only a small fraction of the year but their contribution to the total produced energy may be substantial if the rated power of the electrical equipment is large enough to accommodate that. On the other hand, most of the available wave energy concerns the less energetic sea states that occur most of the time. This raises questions: (i) at which level shall the electrical rated power be fixed, and (ii) how to proceed so that it is not to be exceeded (which could endanger the equipment).

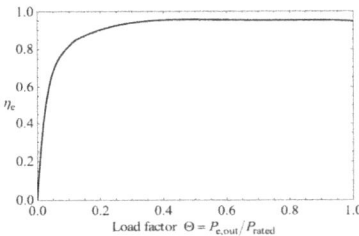

Fig. 10. Electrical efficiency versus electrical load factor for an electrical generator.

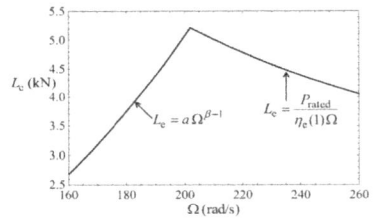

Fig. 11. Control law for OWC converter.

The latter question (ii) is addressed differently depending on the type of wave energy converter. In the case of an OWC device, the power may be constrained by (a) a valve in series or in parallel with the air turbine; or (b) by controlling the electromagnetic torque so that the rated power is not exceeded (Fig. 11), which results in storing the excess energy as kinetic energy of the rotating masses (note that the turbine efficiency drops to very low values at high rotational speeds). Obviously, (b) may be limited by maximum rotational speed constraints (either for the turbine or for the generator). If this is the case, the power available to the turbine has to be reduced. This is done by a mechanical valve to reduce (or simply close) the flow through the turbine. In general, the valve is not fast enough to respond to the power peaks, and its aperture position is changed "from time to time" simply to match the sea state.

The availability of a very fast valve (response time about 0.2 s) in series with the turbine (see Fig. 8) allows what has been called "peak-shaving", i.e. preventing the occurrence of unacceptable power peaks by partially (or if necessary fully) closing the valve, which is left open at the other times. This has been implemented with good results in the biradial turbine tested under real sea conditions while installed at the Mutriku OWC breakwater and later at the Marmok-A-5 spar-buoy OWC, in 2017-19 (European project OPERA), see Fig. 12 [11].

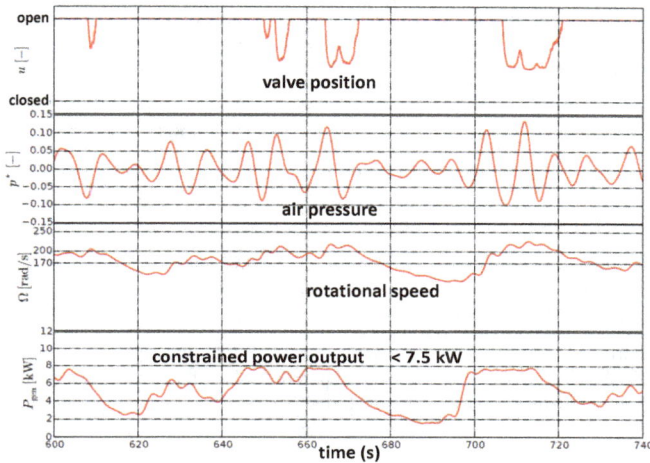

Fig. 12. Measured results of peak-shaving control of one of the OWC converters of the Mutriku breakwater equipped with a biradial turbine and a fast valve [11]. In this test, the electrical generator power was constrained not to exceed 7.5 kW.

References

[1] A.F.O. Falcão, J.C.C. Henriques, Oscillating-water-column wave energy converters and air turbines: A review, Renew. Energy 85 (2016) 1391-1424. https://doi.org/10.1016/j.renene.2015.07.086

[2] G.P.R. Papini, G. Moretti. R. Vertechy, M. Fontana, Control of an oscillating water column wave energy converter based on dielectric elastomer generator, Nonlinear Dyn. 92 (2018) 181-202. https://doi.org/10.1016/j.oceaneng.2006.04.005

[3] P. Boccotti, Comparison between a U-OWC and a conventional OWC, Ocean Eng. 34 (2007) 799-805. https://doi.org/10.1016/j.oceaneng.2006.04.005

[4] A.F.O. Falcão, L.M.C. Gato, E.P.A.S. Nunes, A novel radial self-rectifying air turbine for use in wave energy converters, Renew. Energy 50 (2013) 289-298. https://doi.org/10.1016/j.renene.2012.06.050

[5] P. Benreguig, M. Vicente, A. Dunne, J. Murphy, Modelling approaches of a closed-circuit OWC wave energy converter, J. Marine Sci. Eng. 7 (2019) No. 23. https://doi.org/10.3390/jmse7020023

[6] A.F.O. Falcão, J.C.C. Henriques, The spring-like air compressibility effect in oscillating-water-column wave energy converters: Review and analyses, Renew. Sustain. Energy Rev. 112 (2019) 483-498. https://doi.org/10.1016/j.rser.2019.04.040

[7] J.V. Ringwood, G. Bacelli, F. Fusco, Energy-maximizing control of wave energy converters. The development of control system technology to optimize their operation, IEEEControl Syst. Magaz. 34 (2014) 30-55. https://doi.org/10.1109/MCS.2014.2333253

[8] U.A. Korde, J.V. Ringwood, Hydrodynamic Control of Wave Energy Devices, Cambridge University Press, Cambridge, 2016. https://doi.org/10.1017/CBO9781139942072

[9] J.C.C. Henriques, L.M.C. Gato, A.F.O. Falcão, E. Robles, F.X. Fay, Latching control of a floating oscillating-water-column wave energy converter, Renew. Energy 90 (2016) 229-241. https://doi.org/10.1016/j.renene.2015.12.065

[10] J.C.C. Henriques, A.F.O. Falcão, R.P.F.Gomes, L.M.C. Gato, Latching control of an oscillating water column spar-buoy wave energy converter in regular waves, J. Offshore Mech. Arctic Eng. Trans. ASME, 135 (2013) No. 021902. https://doi.org/10.1115/1.4007595

[11] J.C.C. Henriques, A.A.D. Carrelhas, L.M.C. Gato, A.F.O. Falcão, J.C.C. Portillo, J. Varandas, Peak-shaving control – a new control paradigm for OWC wave energy converters, 13th European Wave Tidal Energy Conf., Naples, Italy, 2019.

Floating Offshore Energy Devices

Materials Research Forum LLC

Materials Research Proceedings **20** (2022) 10-19

https://doi.org/10.21741/9781644901731-2

Cointegration Modelling for Health and Condition Monitoring of Wind Turbines - An Overview

Phong B. Dao[1,a] and Wieslaw J. Staszewski[1,b*]

[1]Department of Robotics and Mechatronics, AGH University of Science and Technology, Al. Mickiewicza 30, 30-059 Kraków, Poland

[a]phongdao@agh.edu.pl, [b]w.j.staszewski@agh.edu.pl

Keywords: Wind Turbine, Condition Monitoring, Fault Detection, Cointegration, Vibration, SCADA

Abstract. The cointegration method has recently attracted a growing interest from scientists and engineers as a promising tool for the development of wind turbine condition monitoring systems. This paper presents a short review of cointegration-based techniques developed for condition monitoring and fault detection of wind turbines. In all reported applications, cointegration residuals are used in control charts for condition monitoring and early failure detection. This is known as the residual-based control chart approach. Vibration signals and SCADA data are typically used with cointegration in these applications. This is due to the fact that vibration-based condition monitoring is one of the most common and effective techniques (used for wind turbines); and the use of SCADA data for condition monitoring and fault detection of wind turbines has become more and more popular in recent years.

Introduction

In recent years, with the fast development of wind power technology, the number and capacity of wind turbines (WTs) have rapidly increased. However, due to the harsh operation environment and time-varying load operation, wind turbines have a high failure rate [1]. It is well known that unexpected failures, especially of large and crucial components, can cause costly repair and excessive downtime. This leads to the increasing of operation and maintenance costs and subsequently the cost of energy. Therefore, it is very important to develop wind turbine monitoring systems that can detect turbine faults at the early stage of fault occurrence. Various condition monitoring techniques have been developed to detect and diagnose abnormalities of WTs, as reviewed in the literature [2-4], such as vibration signal analysis, oil monitoring and analysis, acoustic emission, ultrasonic testing techniques, strain measurement, radiographic inspection, thermography. Another solution – based on the use and analysis of supervisory control and data acquisition (SCADA) data – has been recently developed for early failure detection of wind turbines, as reviewed in [5]. This approach is cost-efficient, readily available, and is beneficial for identifying abnormal components because only key process parameters need to be tracked [1, 5].

Changing environmental and operating conditions of wind turbines are well known to create many difficulties in the signal processing of the measured signals. In particular, wind variations can lead to load variations on the gearbox. Condition monitoring in this case is more challenging and difficult. This implies that monitoring of data trends and removal of undesired effects of environmental and operational variability from wind turbine data are important. Many studies have aimed at developing data analysis/processing methods for effective trend removal, continuous condition monitoring, and reliable abnormal detection of WTs.

Cointegration, a technique originally developed in the econometrics field [6, 7], has recently been introduced to Structural Health Monitoring (SHM) and Condition Monitoring (CM) as a

Floating Offshore Energy Devices
Materials Research Proceedings **20** (2022) 10-19

Materials Research Forum LLC
https://doi.org/10.21741/9781644901731-2

promising data-driven method for the removal of common long-term trends – induced by changing environmental and operational conditions – from the measured data. In essence, the theory of cointegration can be used to combine nonstationary variables to create a stationary combination purged of all common trends in the original data. Therefore, cointegration has been seen as an effective solution to the data normalisation problem in SHM. The main idea behind the use of cointegration for SHM is based on the concept of stationarity and nonstationarity. In a brief description, the variables of interest are cointegrated to create a stationary residual whose stationarity represents normal (or undamaged) condition. Then any departure from stationarity may indicate that monitored processes (or data from monitored structures) are no longer operating under normal condition. Since its first application for the condition monitoring of an industrial distillation unit which was reported in 2009 [8], cointegration has been broadly applied to SHM [9, 10]. These applications have demonstrated that the cointegration process can effectively remove such common long-term trends induced by varying environmental and operational conditions from SHM data. Recently, the cointegration theory has attracted considerable research attention from scientists and engineers worldwide for the development of wind turbine condition monitoring systems. This paper presents a short review of cointegration-based techniques developed for condition monitoring and fault detection of wind turbines in order to demonstrate the state-of-art development of the approach. To the best of the authors knowledge, this issue has not been addressed previously in the literature.

The layout of the paper is organized as follows. Section 2 introduces briefly the cointegration algorithm. Section 3 presents the review and Section 4 provides a discussion on the cointegration-based condition monitoring and fault detection techniques for wind turbines. Finally, the paper is concluded in Section 5.

Cointegration analysis

Consider a time series y_t presented in the form of the first-order Auto-Regressive $AR(1)$ process, which is defined as

$$y_t = \phi y_{t-1} + \varepsilon_t \tag{1}$$

where ε_t is an independent Gaussian white noise process with zero mean, i.e. $\varepsilon_t \sim IWN(0,\sigma^2)$. With different values of the coefficient ϕ, we have three different time series, which are: (1) stationary time series ($|\phi| < 1$); (2) nonstationary time series ($\phi > 1$); and (3) random walk time series ($\phi = 1$).

A random walk time series without a trend is considered as an integrated series of order 1, denoted $I(1)$ [11]. For this time series Eq. (1) yields

$$\Delta y_t = y_t - y_{t-1} = \varepsilon_t \tag{2}$$

Eq. (2) shows that, the first difference of y_t, i.e. $y_t - y_{t-1}$, is a stationary white noise process ε_t. This implies that a nonstationary $I(1)$ time series becomes a stationary $I(0)$ time series after the first difference. In a similar way, a nonstationary $I(2)$ time series requires differencing twice to induce a stationary $I(0)$ time series.

Now, the concept of cointegration can be introduced using a vector Y_t of $I(1)$ time series defined as $Y_t = (y_{1t}, y_{2t}, ..., y_{nt})^T$. One can say that Y_t is linearly cointegrated if there exists a vector $\beta = (\beta_1, \beta_2, ..., \beta_n)^T$ such that

Floating Offshore Energy Devices
Materials Research Proceedings **20** (2022) 10-19

Materials Research Forum LLC
https://doi.org/10.21741/9781644901731-2

$$\beta^T Y_t = \beta_1 y_{1t} + \beta_2 y_{2t} + \cdots + \beta_n y_{nt} \sim I(0) \qquad\qquad (3)$$

In other words, the nonstationary $I(1)$ time series in Y_t are linearly cointegrated if there exists (at least) a linear combination of them that is stationary or has the $I(0)$ status. This linear combination, denoted as $\beta^T Y_t$, is referred to as a cointegration residual or a long-run equilibrium relationship between time series [11]. The vector β is called a cointegrating vector. One can imagine that the cointegration residual ($u_t = \beta^T Y_t$) is created by projecting the vector Y_t on the cointegrating vector β.

A review of cointegration-based approaches to condition monitoring and fault detection of wind turbines

This section presents a short review of recent investigations on cointegration-based condition monitoring and fault detection techniques for wind turbines in order to illustrate the state-of-art development of the approach.

The work in [12-14] presented a novel data analysis/processing method – based on the concept of residual-based control chart – for condition monitoring and fault diagnosis of wind turbines (WTs). The cointegration-based data analysis/processing procedure proposed consists of two stages, i.e. off-line stage and on-line stage, as illustrated in Fig. 1. The main idea of the proposed method relies on the fact that cointegration is a property of some sets of nonstationary time series where a linear combination of these nonstationary series can produce a stationary residual. Then the stationarity (or nonstationarity) of the cointegration residual can be used in a control chart as a potentially effective damage feature. SCADA data – acquired from a WT drivetrain with a nominal power of 2 MW in 30 days under varying environmental and operational conditions – were used to validate the method. Two known problems of the wind turbine (i.e. an abnormal operating state F1 and a gearbox fault F2) were used to illustrate the fault detection ability of the method. The work in [12, 13] used six process parameters of the wind turbine (i.e. wind speed, generator speed, generated power, generator temperature, generator current, gearbox temperature), whereas the investigation in [14] used only the temperature data of gearbox bearing and generator winding. Some selected results of the work in [12] and [14] are shown in Fig. 2 and Fig. 3, respectively. The results have revealed that both studies could effectively monitor the wind turbine and reliably detect abnormal problems with almost the same quality. This confirms that temperature data of the gearbox and generator can provide an early indication of wind turbine faults. Additionally, the use of only gearbox and generator temperature data helps to reduce the number of sensors needed for monitoring the wind turbine. Also, it simplifies the cointegration-based data analysis procedure. What is more, the method proposed in [12-14] has been motivated by the fact of its simplicity and low computational cost in comparison to other commonly used data-mining techniques, e.g., neural network algorithms.

Following the idea of the method developed in [12-14], the work in [15] also used the theory of cointegration for continuously monitoring the operating conditions of wind turbines. First, the optimal combination of different parameters from SCADA data in normal operating condition was determined by using the Johansen's cointegration test and statistical test. Then, the cointegration residuals and stationary threshold boundaries under normal working space were calculated using cointegration analysis. The method was tested using SCADA data of a wind turbine (with nominal power of 2 MW) with known faults. The results have demonstrated that the proposed method can effectively monitor the abnormal state of generator and gearbox, and provide the function of early warning. In addition, the method can monitor several key components of the wind turbine more

Materials Research Forum LLC

https://doi.org/10.21741/9781644901731-2

comprehensively, and avoids the disadvantage of traditional techniques which can only monitor a single parameter.

The work in [16] presented a simulation example of the cointegration-based approach for removing environmental or operational trends from one damage sensitive variable (using a single sensor). This research is meant to be applied for on-line wind turbine gearbox condition monitoring under varying load conditions. Simulation of the dynamic response of a three degree-of-freedom (3-DOF) system to a random excitation was used in the study. In order to introduce effects of imitating environmental or operational conditions, a sinusoidal variation in the stiffness was included in the simulated system. Also, damage was introduced by reducing stiffness parameters. The recursive least square (RLS) algorithm with a forgetting factor $\lambda=0.9$ was used to fit autoregressive (AR) models to the simulated accelerations at the model's masses. The order of the AR models was chosen to be twenty. The method started with estimating twenty coefficients using the RLS method for each acceleration. Subsequently, the first five coefficients produced were used to perform cointegration analysis. A statistical process control X-chart was used for anomaly detection. The results have demonstrated the method's potential to be applied on vibration data measured from a wind turbine transmission system, where cointegration can be adapted as a solution for extracting the load variation influences in the gearbox vibration signals.

Condition monitoring of wind turbine gearboxes based on the cointegration analysis of vibration signals was intensively investigated in [17]. Vibration signals taken from three different points on a Sinovel1500 wind turbine gearbox were choosen as analysis variables. Acceleration sensors were mounted on both high speed and low speed shaft bearing to acquire vibration signals. The authors have discussed that the three vibration signals sampled from the gearbox have similar trends. So, there must be a linear cointegration relationship among these vibration signals. The key idea of the method is to establish a cointegration model of the gearbox in normal condition and then analyse the stability of residuals calculated by the cointegration model. Once a gearbox failure occurs, vibration features of the testing point which is close to the position of failure will be changed. Consequently, the cointegration relationship is broken and the stability of cointegration residuals changes accordingly. This work also used statistical process control to set the thresholds of residuals as the failure warning level. Through the simulation analysis of gearbox fault data, the results verify the effectiveness of cointegration in monitoring condition of wind turbine gearbox. Selected results of this work are shown in Fig. 4.

A cointegration-based monitoring method for rolling bearings working in time-varying operational conditions was recently developed in [18]. The proposed method was applied to vibration signals measured on an experimental bearing test rig. The signals – acquired during run-up condition – were first decomposed into zero-mean modes called intrinsic mode functions using the improved ensemble empirical mode decomposition method. Next, cointegration analysis was applied to the intrinsic mode functions to extract stationary residuals. The feature vectors were then created by applying the Teager-Kaiser energy operator to the stationary residuals. Finally, the feature vectors of the healthy bearing signals were utilised to construct a separating hyperplane using the one-class support vector machine method. The results confirmed that the method could successfully distinguish between healthy and faulty bearings even if the shaft speed changes considerably.

An interesting application of cointegration to analyse vibration signals for local damage detection in gearboxes was presented in [19]. The work started with the assumption that the correlation of given vibration signal is periodic and its period can be measured. Then, signal was restructured and divided into sub-signals according to the discovered period. Next, sub-signals

Floating Offshore Energy Devices   Materials Research Forum LLC
Materials Research Proceedings **20** (2022) 10-19   https://doi.org/10.21741/9781644901731-2

were checked if they are integrated and the cointegrating vector was calculated by using the least squares method. Finally, in order to test if the cointegrating vector corresponds to healthy or damaged gearbox the authors examined whether it exhibits random (chaotic) behaviour by using the Wald-Wolfowitz test for randomness. The proposed methodology was validated using simulated vibration signals and real data from a two-stage gearbox (with first stage being conical and second cylindrical) used in mining industry. Based on the analysis of cointegrating vectors the damaged gearbox could be detected.

In [20], the authors have explored the use of cointegration in detection of structural damage in the blade of an operating Vestas V27 wind turbine under the effect of certain environmental and operational variabilities (EOVs). The experimental campaign included a measurement period of 3.5 months, in which the blade in question was instrumented with 11 piezoelectric accelerometers (distributed along the leading and trailing edge). The wind turbine was analysed in a healthy/reference state and three damaged scenarios where a trailing edge opening was introduced gradually to the instrumented blade with increasing size from 15 cm long, to 30 cm long, and finally 45 cm long. In addition to acceleration measurements in the different structural states, the study recorded the varying environmental and operational conditions (including wind speed and direction at different altitudes, ambient temperature, atmospheric pressure and precipitation) over the 3.5-months period. The Q-statistics was employed as the damage metric to quantify the discordance between the statistical baseline representing the healthy structural state and realizations from the potentially damaged state. The results have demonstrated that cointegration can be used to successfully detect the introduced damages under conditions not allowing for direct discrimination between damage and EOVs.

It should be noted here that all applications – reported in [8-10] for SHM systems and in [12-20] for wind turbine condition monitoring – have used the linear cointegration theory that was originally developed in [6, 7] and intimately connected with the concept of linear error correction models. However, it is well known that response signals (e.g. Lamb waves, vibration data, SCADA data) acquired from engineering structures or wind energy systems often exhibit not only nonstationarity, but also nonlinear behaviour. Moreover, operational and environmental trends are typically believed to be nonlinearly related with response data used for damage detection or condition monitoring. If this is the case then the conventional linear cointegration theory might be no longer suitable for structural damage detection as well as process condition monitoring and therefore nonlinear cointegration approaches are highly needed.

The work in [21] brought the concept of nonlinear cointegration to SHM. However, a major problem was observed, that is, the variance of cointegration residuals (calculated for a healthy structure) increased with time, although cointegrated variables were mean stationary. This behaviour – known in mathematics as the heteroscedasticity – implied that strictly stationary cointegration residuals could not be obtained. When a cointegration residual with unstable variance characteristics is used in a control chart (i.e. statistical process control) for condition monitoring of a wind turbine, it is not possible to identify accurately that whether a gearbox failure occurs when the residual exceeds a threshold. So, it is clear that reliable condition monitoring methods for WTs based on nonlinear cointegration would require homoscedastic cointegration residuals (i.e. strictly stationary residuals) to prevent false diagnosis results. Recently, the work in [22, 23] has investigated a new approach to nonlinear cointegration, with applications towards SHM and wind turbine condition monitoring – which could solve the problems of heteroscedasticity and nonlinear trend removal. As a result, an approximately homoscedastic nonlinear cointegration method has been proposed for the removal of undesired (environmental, operational or measurement) trends from SHM data in general and wind turbine SCADA data in particular. The method has been successfully applied for condition monitoring and fault detection of a wind turbine drivetrain with a nominal power of 2 MW in the presence of nonlinearity between operational parameters.

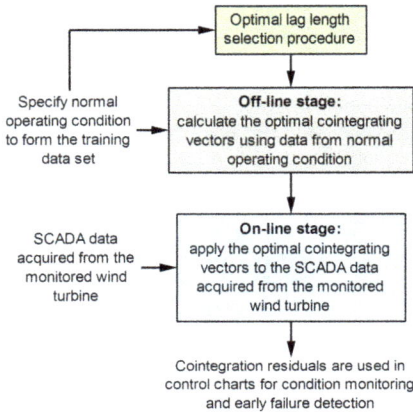

Fig. 1 Cointegration-based data analysis procedure for condition monitoring of wind turbines using SCADA data [12].

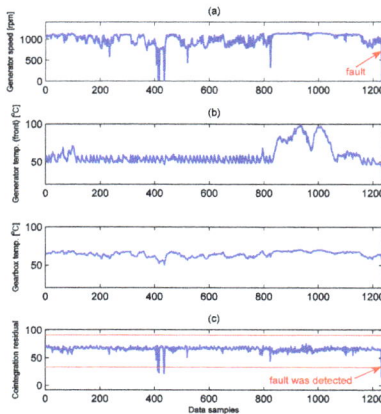

Fig. 2 Condition monitoring and fault detection of the wind turbine using multiple process parameters [12]: (a) abnormal operating state (F1) and gearbox fault (F2); (b) monitoring of F1 and F2 using the 1st and 5th cointegration residuals in control charts.

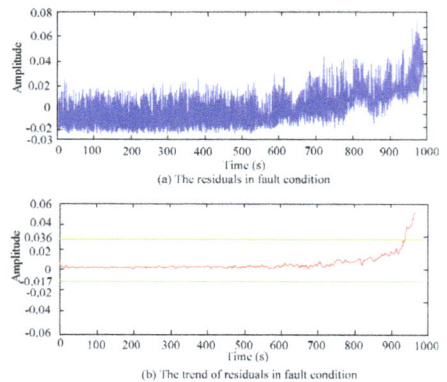

Fig. 3 Condition monitoring of the wind turbine using only the temperature data of gearbox and generator [14].

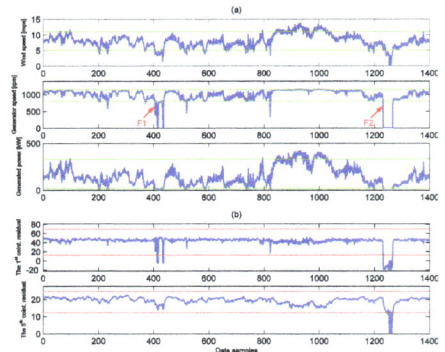

Fig. 4 Monitoring of the wind turbine conditions using cointegration [17]: (a) the residual in fault condition; (b) the residual trend in fault condition.

Floating Offshore Energy Devices
Materials Research Proceedings **20** (2022) 10-19

Materials Research Forum LLC
https://doi.org/10.21741/9781644901731-2

Discussion

Some important remarks and a comprehensive comparison between the cointegration-based method and other relevant approaches for condition monitoring and fault detection of wind turbines are presented in the following.

First of all, it should be noted that the major idea of the cointegration-based condition monitoring and fault detection techniques for wind turbines [12-23], as reviewed in Section 3, is basically relied on the well-known control chart approach, which is one of the primary techniques of statistical process control. Basically, control charts plot the quality characteristic as a function of the sample number. The charts have lower and upper control limits, which are computed from the samples recorded when the process is assumed to be in control. When abnormal sources of variability are present, sample statistics will plot outside the control limits and an alarm signal will be produced. An advantage of control charts is that they can be automated for on-line structural health monitoring.

Second, condition monitoring systems of wind turbines, as reported in Section 3, have employed the cointegration technique for either vibration data [16-20] or SCADA data [12-15, 22, 23]. It is due to the fact that vibration signals of gearbox contain a large number of operating condition information. Hence, it is common to use vibration signals for fault prediction and diagnosis. Regarding SCADA-based approaches, since standard SCADA systems have been installed in the majority of utility-scale WTs for system control, data logging and performance monitoring so that the data needed for analysis is readily available and no additional hardware and sensors are required when developing a SCADA-based condition monitoring (CM) system [5, 12]. Hence, this is a potentially low cost solution. In addition, SCADA-based CM systems can be designed to operate in on-line or off-line mode. Because of these advantages, SCADA data have been used with cointegration to develop monitoring systems for WTs. It is suggested that if gearbox vibration signals of a wind turbine are combined with its SCADA data for cointegration analysis, earlier fault prediction can be achieved with high accuracy.

Next, it should be mentioned that regression analysis can be used for condition monitoring of wind power systems, as illustrated in [24]. However, cointegration analysis has been used in [12-23] instead of other regression techniques is due to two main reasons: (1) to avoid the problem of spurious regression; and (2) to actively deal with the undesired effect of environmental and operational conditions in the analysed data. The former has been discussed broadly in the econometrics literature [25]. The problem arises when standard regression analysis fails while dealing with nonstationary variables, leading to spurious regressions that suggest relationships even when there are none. For example, if two time series show monotonic trends, even if the trends are not causally related, ordinary least-squares (OLS) regression will potentially find a spurious relationship. The later relates to the capability of cointegartion analysis for removing undesired effects of environmental and operational variability from wind turbine SCADA data (SHM data in general), while still maintaining sensitivity of cointegration residuals to faults, structural damage, or abnormal problems. This process is known as data normalisation.

Finally, in comparison with typical data-mining algorithms, such as neural network (NN), support vector machines, adaptive neuro-fuzzy interference systems (ANFIS), decision tree learning, or naive Bayes classifier, cointegration-based condition monitoring algorithms are simpler and requires much less computational resources. For example, in the study [12-14], the calculation of cointegrating vectors in the off-line stage takes only few seconds on a normal computer. Then, the cointegration residual is obtained through projecting the SCADA data – acquired from the monitored WT under regular working phase for producing electricity – on the

16

resulting cointegrating vectors. This is done simply by multiplying a vector of time series variables by one cointegrating vector to form one cointegration residual, or multiplying a matrix of time series variables by   cointegrating vectors to obtain   cointegration residuals. This computation process can be promptly executed in real-time manner on a computer-based condition monitoring system, thereby providing a simple on-line condition monitoring solution for wind turbines. Furthermore, cointegration can be used in practice to monitor a wind turbine system without the need of analysing many nonstationary variables. Through monitoring a cointegration residual, one can achieve the objective of simultaneous monitoring of multiple nonstationary variables [12, 16].

Conclusions

This paper have reviewed recent investigations on cointegration-based condition monitoring and fault diagnosis techniques for wind turbines. First of all, it is observed that all reported applications have used cointegration residuals in control charts for condition monitoring and early failure detection. This is known as the residual-based control chart approach. Second, only vibration data and SCADA data have been used with cointegration in these applications so far. This is due to the fact that vibration signals are the most common condition monitoring signals and SCADA-based condition monitoring has become more and more popular in recent years.

An important conclusion is that the cointegration-based techniques can automatically interpret and analyse a large amount of low-sampling rate SCADA data and enables a transition from a singular process parameter analysis to automatic interpretation and analysis of a large number of process parameters. Moreover, simplicity and fast computation are the major advantages of cointegration-based techniques, if comparted with other common techniques (such as NN-based and ANFIS-based algorithms). Hence, the cointegration-based condition monitoring algorithm for wind turbines using vibration signals and SCADA data can be computed on-line and deployed on a computer for real-time condition monitoring applications.

Furthermore, the use of cointegration can remove, compensate, or at least, mitigate the effect of environmental and operational variability in vibration and SCADA data used for condition monitoring and fault detection of wind turbines.

References

[1] A. Kusiak, W. Li, The prediction and diagnosis of wind turbine faults, Renew. Energy. 36 (2011) 16-23. https://doi.org/10.1016/j.renene.2010.05.014

[2] Z. Hameed, Y.S. Hong, Y.M. Cho, S.H. Ahn, C.K. Song, Condition monitoring and fault detection of wind turbines and related algorithms: a review, Renew. Sust. Energ. Rev. 13 (2009) 1-39. https://doi.org/10.1016/j.rser.2007.05.008

[3] Y. Amirat, M.E.H. Benbouzid, E. Al-Ahmar, B. Bensaker, S. Turri, A brief status on condition monitoring and fault diagnosis in wind energy conversation systems, Renew. Sust. Energ. Rev. 13 (2009) 2629-2636. https://doi.org/10.1016/j.rser.2009.06.031

[4] F.P. Garcia Marquez, A.M. Tobias, J.M. Pinar Perez, M. Papaelias, Condition monitoring of wind turbines: techniques and methods, Renew. Energy. 46 (2012) 169-178. https://doi.org/10.1016/j.renene.2012.03.003

[5] J. Tautz-Weinert, S.J. Watson, Using SCADA data for wind turbine condition monitoring – a review, IET Renewable Power Gener. 11 (2017) 382-394. https://doi.org/10.1049/iet-rpg.2016.0248

Materials Research Forum LLC
https://doi.org/10.21741/9781644901731-2

[6] R.F. Engle, C.W.J. Granger, Cointegration and error-correction: representation, estimation and testing, Econometrica, 55 (1987) 251-276. https://doi.org/10.2307/1913236

[7] S. Johansen, Statistical analysis of cointegration vectors, Journal of Economic Dynamics and Control, 12 (1988) 231-254. https://doi.org/10.1016/0165-1889(88)90041-3

[8] Q. Chen, U. Kruger, A.Y.T. Leung, Cointegration testing method for monitoring non-stationary processes, Ind. Eng. Chem. Res. 48 (2009) 3533-3543. https://doi.org/10.1021/ie801611s

[9] E.J. Cross, K. Worden, Q. Chen, Cointegration: a novel approach for the removal of environmental trends in structural health monitoring data, Proceedings of the Royal Society A. 467 (2011) 2712-2732. https://doi.org/10.1098/rspa.2011.0023

[10] P.B. Dao, W.J. Staszewski, Cointegration approach for temperature effect compensation in Lamb wave based damage detection, Smart Mater. Struct. 22 (2013) 095002. https://doi.org/10.1088/0964-1726/22/9/095002

[11] E. Zivot, J. Wang, Modeling Financial Time Series with S-PLUS, 2nd ed., New York (NY): Springer, 2006.

[12] P.B. Dao, W.J. Staszewski, T. Barszcz, T. Uhl, Condition monitoring and fault detection in wind turbines based on cointegration analysis of SCADA data, Renew. Energy. 116 (2018) 107-122. https://doi.org/10.1016/j.renene.2017.06.089

[13] P.B. Dao, W.J. Staszewski, T. Uhl, Operational condition monitoring of wind turbines using cointegration method, in: A. Timofiejczuk, F. Chaari, R. Zimroz, W. Bartelmus, M. Haddar (Eds.), Advances in Condition Monitoring of Machinery in Non-Stationary Operations, Applied Condition Monitoring, vol. 9, chapter 21, Springer, Cham, 2018, pp. 223-233. https://doi.org/10.1007/978-3-319-61927-9_21

[14] P.B. Dao, Condition monitoring of wind turbines based on cointegration analysis of gearbox and generator temperature data, Diagnostyka. 19 (2018) 63-71. https://doi.org/10.29354/diag/81298

[15] X. Sun, D. Xue, R. Li, X. Li, L. Cui, X. Zhang, W. Wu, Research on condition monitoring of key components in wind turbine based on cointegration analysis, IOP Conf. Ser.: Mater. Sci. Eng. 575 (2019), Article ID 012015. https://doi.org/10.1088/1757-899X/575/1/012015

[16] I. Antoniadou, E.J. Cross, K. Worden, Cointegration for the removal of environmental and operational effects using a single sensor, in: Proceedings of the 9th International Workshop on Structural Health Monitoring, vols. 1 and 2, 2013, pp. 2400-2406.

[17] H. Zhao, H. Liu, H. Ren, H. Liu, The condition monitoring of wind turbine gearbox based on cointegration, in: Proceedings of 2016 IEEE International Conference on Power System Technology (POWERCON), 2016, pp. 1-6. https://doi.org/10.1109/POWERCON.2016.7753906

[18] A.A. Tabrizi, H. Al-Bugharbee, I. Trendafilova, L. Garibaldi, A cointegration-based monitoring method for rolling bearings working in time-varying operational conditions, Meccanica. 52 (2017) 1201-1217. https://doi.org/10.1007/s11012-016-0451-x

Materials Research Forum LLC

https://doi.org/10.21741/9781644901731-2

[19] A. Michalak, J. Wodecki, A. Wyłomańska, R. Zimroz, Application of cointegration to vibration signal for local damage detection in gearboxes, Appl. Acoust. 144 (2017) 4-10. https://doi.org/10.1016/j.apacoust.2017.08.024

[20] B.A. Qadri, M.D. Ulriksen, L. Damkilde, D. Tcherniak, Cointegration for detecting structural blade damage in an operating wind turbine: an experimental study, in: S. Pakzad (Ed.), Dynamics of Civil Structures, Conference Proceedings of the Society for Experimental Mechanics Series, vol. 2, chapter 23, Springer, Cham, 2020, pp. 173-180. https://doi.org/10.1007/978-3-030-12115-0_23

[21] E.J. Cross, K. Worden, Approaches to nonlinear cointegration with a view towards applications in SHM, J. Phys. Conf. Ser. 305 (2011) 012069. https://doi.org/10.1088/1742-6596/305/1/012069

[22] K. Zolna, P.B. Dao, W.J. Staszewski, T. Barszcz, Nonlinear cointegration approach for condition monitoring of wind turbines, Math. Prob. Eng. 2015 (2015), Article ID 978156. https://doi.org/10.1155/2015/978156

[23] K. Zolna, P.B. Dao, W.J. Staszewski, T. Barszcz, Towards homoscedastic nonlinear cointegration for structural health monitoring, Mech. Syst. Sig. Process. 75 (2016) 94-108. https://doi.org/10.1016/j.ymssp.2015.12.014

[24] N. Yampikulsakul, E. Byon, S. Huang, S. Sheng, M. You, Condition monitoring of wind power system with nonparametric regression analysis, IEEE Trans. Energy Convers. 29 (2014) 288-299. https://doi.org/10.1109/TEC.2013.2295301

[25] P.C.B. Phillips, Understanding spurious regressions in econometrics, J. Econom. 33 (1986) 311-340. https://doi.org/10.1016/0304-4076(86)90001-1

Floating Offshore Energy Devices
Materials Research Proceedings **20** (2022) 20-25

Materials Research Forum LLC
https://doi.org/10.21741/9781644901731-3

On the Coriolis Effect for Internal Ocean Waves

Rossen Ivanov

School of Mathematical Sciences, TU Dublin, City Campus, Grangegorman,
Dublin 7, Ireland

rossen.ivanov@TUDublin.ie

Keywords: Internal Waves, Hamiltonian, KdV Equation, Boussinesq Equation, Ostrovsky Equation, Tidal Motion

Abstract. A derivation of the Ostrovsky equation for internal waves with methods of the Hamiltonian water wave dynamics is presented. The internal wave formed at a pycnocline or thermocline in the ocean is influenced by the Coriolis force of the Earth's rotation. The Ostrovsky equation arises in the long waves and small amplitude approximation and for certain geophysical scales of the physical variables.

Introduction

The internal ocean waves could have a significant impact on offshore engineering structures, such as oil platforms in the oceans as well as stationary tubes for oil and gas transportation stretching along the ocean shelf slope [1]. Builders of underwater constructions in equatorial districts, for example, experience the influence of huge underwater internal waves and strong surface flows, which are interfering with their work activities.

The internal waves often are driven by tidal motion. The open water exploitation of tidal and wave power is under current considerations. It has been estimated globally that 180 TWh of economically accessible tidal energy is available. However, due to geographical, technical, and environmental constraints, only a fraction of this could be captured in practical terms [5].

The pattern of the ocean movement around the points of zero tidal wave amplitude (amphidromic point) is due to the Coriolis effect. Therefore there are deep interrelations between the tidal motion, internal waves and Coriolis forces that deserve detailed studies, since these are of potential practical significance.

In this work we examine the Coriolis effect on the internal wave propagation following the idea of *nearly* Hamiltonian approach, developed in series of previous papers like [6, 4] and [3] and generalising the Hamiltonian approach of Zakharov [14].

A mass of moving air or water subject only to the Coriolis force travels in a circular trajectory called an *inertial circle*, for the atmosphere see the illustration on Fig. 1(a). For ocean waves the Coriolis Effect is not so pronounced, nevertheless it affects the wave propagation. For ocean waves of large magnitude, the viscosity does not play an essential role and can be neglected, so effectively the fluid dynamics is govern by Euler's equation.

Internal Waves with Coriolis force - the Setup

The Euler equation with included Coriolis force is

$$V_t + (V \cdot \nabla)V + 2\vec{\omega} \times V = -\frac{1}{\rho}\nabla p + \vec{g} \tag{1}$$

Floating Offshore Energy Devices | Materials Research Forum LLC
Materials Research Proceedings **20** (2022) 20-25 | https://doi.org/10.21741/9781644901731-3

where the velocity vector field $V = (u, v, w)$ is presented through its components in a local coordinate system where the geophysical axis x is oriented to the East, the y axis is pointing to the North and the z axis is vertical to the Earth's surface, $\vec{g} = (0,0,-g)$ is the Earth's gravity acceleration. In addition we have incompressibility, i.e. div $V = 0$. p is the pressure in the fluid. The Earth's angular velocity at latitude in this system is $\vec{\omega} = \omega\,(0, \cos\theta, \sin\theta)$, $\omega = 7.3 \times 10^{-5}$ rad/s. Introducing the parameters $f = 2\omega\sin\theta$ and $r = 2\omega\cos\theta$ we have

$$2\vec{\omega} \times V = (rw - fv, fu, -ru).$$

For Equatorial motion $\theta = 0$ and $f = 0$ so there are no forces acting in the y-direction. Moreover, the Coriolis forces are supporting the fluid to move along the Equator (in the x-direction), so that its motion remains two-dimensional. Such situation with internal equatorial waves and currents is studied in [4].

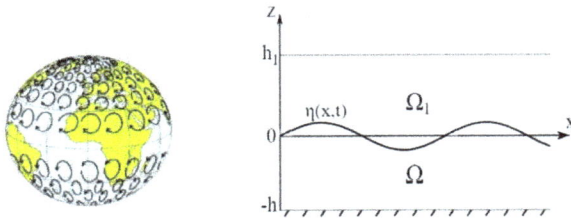

Fig. 1: (a) Left: Coriolis forces in the atmosphere. Schematic representation of inertial circles of air masses in the absence of other forces. Source: Wikipedia; (b) Right: System with an internal wave. The fluid domain Ω contains fluid of higher density. The pycnocline/thermocline separates the two fluid domains Ω and Ω_1. The function $\eta(x, t)$ describes the elevation of the internal wave.

We are going to consider now for example $\theta > 0$. In addition we assume that the fluid motion is irrotational (i.e. absence of currents and vorticity), apart from the global rotation caused by the Coriolis forces. In this approximation the velocity field is potential, i.e. $V = \nabla\boldsymbol{\varphi}(x, y, z, t)$ and the Coriolis Effect will be presented as a perturbation to the potential motion. The governing equations (1) acquire the form:

$$\left(\varphi_t + \frac{|\nabla\varphi|^2}{2} + \frac{p}{\rho} + gz\right)_x + r\varphi_z - f\varphi_y = 0,$$

$$\left(\varphi_t + \frac{|\nabla\varphi|^2}{2} + \frac{p}{\rho} + gz\right)_y + f\varphi_x = 0, \tag{2}$$

$$\left(\varphi_t + \frac{|\nabla\varphi|^2}{2} + \frac{p}{\rho} + gz\right)_z - r\varphi_x = 0,$$

where p is the pressure in the fluid. The internal waves are illustrated on Fig. 1(b). For fixed y the system is bounded at the bottom by an impermeable flatbed and is considered as being bounded on the surface by an assumption of absence of surface motion. The domains Ω and Ω_1 are defined with values associated with each domain using corresponding respective subscript notation. Also, subscript c (implying common interface) will be used to denote evaluation on the internal wave $z = \eta\,(x,\ t)$. Propagation of the internal wave is assumed to be in the positive x-direction which is considered to be eastward. The function $\eta\,(x,\ t)$ describes the elevation of the internal wave with the spatial mean of $\eta(x,t)$ assumed to be zero. The system is considered incompressible with ρ and ρ_1 being the respective constant densities of the lower and upper media and stability is given by the immiscibility condition $\rho \gg \rho_1$. For long internal waves the parameter $\delta = \frac{h}{\lambda} \ll 1$ and φ is a small quantity of order δ, see for example [4]. The terms proportional to r lead to a very small correction (0.01%) of the wave propagation speed c_0 in the x-direction, for the feasible values of the parameters (see for example the calculations in [4]) and thus they can be neglected. The motion in the y direction is very slow in comparison to the wave propagation in the x-direction, therefore in leading order we have $p = p(x, z)$ and we can use the second equation in Eq. (2) in linear approximation to exclude the y dependence, $\varphi_{ty} + f\varphi_x = 0$ giving formally

$$\varphi_y = -f\partial_t^{-1}\varphi_x.$$

Assuming further that f is of order $\delta^2 \ll 1$ and noting that the ∂_x operator with an eigenvalue $k = \frac{2\pi}{\lambda}$ is also of order δ, for compatible time-scales $\partial_t \sim \delta$ thus we see that the y-derivative of φ is $\varphi_y \sim \delta\varphi_x$ (more details about the scales could be found in [4]). The first equation in (2) gives the following generalisation of the Bernoulli equation:

$$\varphi_t + \frac{|\nabla\varphi|^2}{2} + \frac{p}{\rho} + gz + f^2\partial_t^{-1}\varphi = 0.$$

Therefore in the nonlinear contribution $|\nabla\varphi|^2 = \varphi_x^2 + \varphi_y^2 + \varphi_z^2$ the first term is $\sim\delta^4$ already small and the second is $\sim\delta^6$ (much smaller) and could be neglected in comparison to φ_x^2 giving $|\nabla\varphi|^2 \approx \varphi_x^2 + \varphi_z^2$ or

$$\varphi_t + \frac{\varphi_x^2 + \varphi_z^2}{2} + \frac{p(x,z)}{\rho} + gz + f^2\partial_t^{-1}\varphi = 0. \qquad (3)$$

We can proceed now with this effectively *(2+1)*-dimensional equation for the x and z dependent variables, considering y fixed, since there are no y-derivatives.

(Nearly) Hamiltonian representation of the internal wave dynamics
The propagation of the internal wave is assumed to be in the positive x-direction which is considered to be *eastward*. At $z = \eta\,(x,\ t)$ we have $p(x, \eta, t) = p_1(x, \eta, t)$ and therefore Eq. (3) gives the Bernoulli condition

$$\rho\left((\varphi_t)_c + \frac{(\varphi_x^2 + \varphi_z^2)_c}{2} + g\eta + f^2(\partial_t^{-1}\varphi)_c\right) = \rho_1\left((\varphi_{1,t})_c + \frac{(\varphi_{1,x}^2 + \varphi_{1,z}^2)_c}{2} + g\eta + f^2(\partial_t^{-1}\varphi_1)_c\right)$$

The last equation suggests the introduction of the variable $\xi(x,t) = (\rho\varphi - \rho_1\varphi_1)_c$. Indeed, following [6, 4] this equation can be written in the nearly Hamiltonian form

$$\xi_t = -\frac{\delta H_0}{\delta \eta} - (\delta^2)f^2(\partial_t^{-1}(\rho\varphi - \rho_1\varphi_1))_c \tag{4}$$

for the Hamiltonian (expansion with respect to the scale parameter δ, following the leading order; $D = -i\partial_x \sim \delta, \eta \sim \delta^2$)

$$H_0(\xi, \eta) = \frac{1}{2}\int_{\mathbb{R}} \xi D(\alpha_1 + \delta^2(\alpha_3\eta - \alpha_2 D^2))D\xi dx + \frac{1}{2}g(\rho - \rho_1)\int_{\mathbb{R}} \eta^2 dx \tag{5}$$

where

$$\alpha_1 = \frac{hh_1}{\rho_1 h + \rho h_1}, \quad \alpha_2 = \frac{h^2 h_1^2(\rho h + \rho_1 h_1)}{3(\rho_1 h + \rho h_1)^2}, \quad \alpha_3 = \frac{\rho h_1^2 - \rho_1 h^2}{(\rho_1 h + \rho h_1)^2}. \tag{6}$$

The kinematic boundary condition on the interface leads to the second equation,

$$\eta_t = \frac{\delta H_0}{\delta \xi} \tag{7}$$

so that Eq. (4) and Eq. (7) represent the *nearly* Hamiltonian formulation of the internal wave dynamics in the long-wave -small amplitude approximation.

Boussinesq and KdV type approximations. Ostrovsky equation
Introducing the variable $\tilde{u} = \xi_x$ one can verify by a simple computation that $\partial_x(\partial_t^{-1}(\rho\varphi - \rho_1\varphi_1))_c = \partial_t^{-1}\tilde{u} +$ smaller order terms. Then the equations (4) and (7) in terms of η and $\tilde{u} = \xi_x$ are

$$\eta_t + \alpha_1\tilde{u}_x + \delta^2\alpha_2\tilde{u}_{xxx} + \delta^2\alpha_3(\eta\tilde{u})_x = 0, \tag{8}$$

$$\tilde{u}_t + g(\rho - \rho_1)\eta_x + \delta^2\alpha_3\tilde{u}\tilde{u}_x + \delta^2 f^2(\partial_t^{-1}\tilde{u}) = 0. \tag{9}$$

In leading order $\eta_t + \alpha_1\tilde{u}_x = 0$, $\tilde{u}_t + g(\rho - \rho_1)\eta_x = 0$ or
$\eta_{tt} = -\alpha_1\tilde{u}_{xt} = g\alpha_1(\rho - \rho_1)\eta_{xx}$, $\eta_{tt} - g\alpha_1(\rho - \rho_1)\eta_{xx} = 0$, which is the wave equation for η giving the wave speed

$$c_0 = \pm\sqrt{g\alpha_1(\rho - \rho_1)}.$$

For an observer, moving with the flow, i.e. there are left- (minus sign) and right-running (+ sign) waves. Moreover, in the leading approximation, for linear waves, the functions depend on the characteristic variable $x - c_0 t$, therefore $\tilde{u} = \frac{c_0}{\alpha_1}\eta$. In the next order approximations with respect to the scale parameter δ obviously

$$\tilde{u} = \frac{c_0}{\alpha_1}\eta + \delta^2(\cdots). \tag{10}$$

Floating Offshore Energy Devices
Materials Research Proceedings **20** (2022) 20-25

Materials Research Forum LLC
https://doi.org/10.21741/9781644901731-3

however we will not need this explicitly, see for example [10]. Differentiating Eq. (8) with respect to t

$$\eta_{tt} + \alpha_1 \tilde{u}_{tx} + \delta^2 \alpha_2 \tilde{u}_{txxx} + \delta^2 \alpha_3 (\eta_t \tilde{u} + \eta \tilde{u}_t)_x = 0,$$

and substituting in it $\tilde{u}_t$ from Eq. (9), $\tilde{u}$ from Eq. (10) and $\eta_t = -\alpha_1 \tilde{u}_x + \delta^2(...)$ where necessary, neglecting δ^4 terms, we obtain the following generalised Boussinesq equation for η:

$$\eta_{tt} - c_0^2 \eta_{xx} - \delta^2 \frac{3\alpha_3 c_0^2}{2\alpha_1} (\eta^2)_{xx} - \delta^2 \frac{\alpha_2 c_0^2}{\alpha_1} \eta_{xxxx} + \delta^2 f^2 \eta = 0. \tag{11}$$

The dispersion law of this equation is $\widetilde{\omega}^2(k) = c_0^2 k^2 - \delta^2 \frac{\alpha_2 c_0^2}{\alpha_1} k^4 + \delta^2 f^2$ or approximately

$$\widetilde{\omega}(k) = c_0 k - \delta^2 \frac{\alpha_2 c_0^2}{2\alpha_1} k^3 + \delta^2 \frac{f^2}{2kc_0}. \tag{12}$$

Furthermore, a generalised KdV type equation of the form

$$\eta_t + c_0 \eta_x + \delta^2 a \eta_{xxx} + \delta^2 b(\eta^2)_x + \delta^2 n f^2 \partial_x^{-1} \eta = 0 \tag{13}$$

for some constants a, b, n (yet unknown) could be obtained from Eq. (11). Indeed, differentiating the above equation with respect to t we have

$$\eta_{tt} + c_0 \eta_{xt} + \delta^2 a \eta_{txxx} + \delta^2 b(\eta^2)_{xt} + \delta^2 n f^2 \partial_x^{-1} \eta_t = 0 \tag{14}$$

in which we substitute η_t from Eq. (13) to obtain (neglecting δ^4 terms)

$$\eta_{tt} - c_0^2 \eta_{xx} - \delta^2 2bc_0(\eta^2)_{xx} - \delta^2 ac_0 \eta_{xxxx} - \delta^2 2nc_0 f^2 \eta = 0.$$

The comparison with Eq. (11) gives

$$a = \frac{\alpha_2 c_0}{2\alpha_1}, \qquad b = \frac{3\alpha_3 c_0}{4\alpha_1}, \qquad n = -\frac{1}{2c_0}.$$

Then finally the KdV-type equation acquires the form

$$\eta_t + c_0 \eta_x + \delta^2 \frac{\alpha_2 c_0}{2\alpha_1} \eta_{xxx} + \delta^2 \frac{3\alpha_3 c_0}{4\alpha_1} (\eta^2)_x = \delta^2 \frac{f^2}{2c_0} \partial_x^{-1} \eta.$$

This is also known as Ostrovsky equation [12]. Note that the dispersion law of the Ostrovsky equation is like in Eq. (12).

For surface waves the derivation is analogous, only the Hamiltonian H_0 is the KdV Hamiltonian for surface waves. The derivation directly from Euler's equations could be found in Leonov's paper [11]. The Ostrovsky equation itself is Hamiltonian and possesses three conservation laws, however it is not bi-Hamiltonian and it is not integrable by the Inverse Scattering Method [2]. Solutions from perturbations of the KdV solitons can be derived in principle, although this is technically difficult, see for example [8] and the references therein. Various other aspects of the equation have been studied extensively by now in numerous works, see for example [9, 13] and the references therein.

References

[1] V.V. Bulatov, Y.V. Vladimirov, Fundamental problems of internal gravity waves dynamics in ocean, Journal of Basic & Applied Sciences 9 (2013) 69-81. https://doi.org/10.6000/1927-5129.2013.09.12

[2] R. Choudhury, R.I. Ivanov and Y. Liu, Hamiltonian formulation, nonintegrability and local bifurcations for the Ostrovsky equation, Chaos, Solitons and Fractals 34 (2007) 544-550. https://doi.org/10.1016/j.chaos.2006.03.057

[3] A. Compelli, Hamiltonian approach to the modelling of internal geophysical waves with vorticity, Monatsh. Math. 179(4) (2016), 509-521. https://doi.org/10.1007/s00605-014-0724-1

[4] A. Compelli, R.I. Ivanov, The dynamics of at surface internal geophysical waves with currents, Journal of Mathematical Fluid Mechanics, 19, (2017) 329-344. https://doi.org/10.1007/s00021-016-0283-4

[5] E. Coyle, B. Basu, J. Blackledge and W. Grimson, Harnessing Nature: Wind, Hydro, Wave, Tidal, and Geothermal Energy, in: *Understanding the Global Energy Crisis* by E. Coyle and R. Simmons (Eds.), Purdue University Press. (2014), URL: https://www.jstor.org/stable/j.ctt6wq56p.9

[6] A. Constantin, R.I. Ivanov, E.M. Prodanov, Nearly-Hamiltonian structure for water waves with constant vorticity, J. Math. Fluid Mech. 9 (2007), 1-14; arXiv:math-ph/0610014

[7] W. Craig, P. Guyenne, H. Kalisch, Hamiltonian long wave expansions for free surfaces and interfaces, Comm. Pure Appl. Math. 24 (2005), 1587-1641. https://doi.org/10.1002/cpa.20098

[8] G. Grahovski, R. Ivanov, Generalised Fourier transform and perturbations to soliton equations. Discrete Contin. Dyn. Syst. Ser. B 12 (2009), 579-595. https://doi.org/10.3934/dcdsb.2009.12.579

[9] R. Grimshaw, L. Ostrovsky, V. Shrira, Y. Stepanyants, Long nonlinear surface and internal waves in a rotating ocean, Surveys Geophys. 19 (1998) 289-338. https://doi.org/10.1023/A:1006587919935

[10] R.S. Johnson, Camassa-Holm, Korteweg-de Vries and related models for water waves, J. Fluid. Mech. 457 (2002), 63-82. https://doi.org/10.1017/S0022112001007224

[11] A.I. Leonov, The effect of the Earth's rotation on the propagation of weak non-linear surface and internal long oceanic waves, Ann. NY Acad. Sci. 373 (1981) 150-159. https://doi.org/10.1111/j.1749-6632.1981.tb51140.x

[12] L.A. Ostrovsky, Nonlinear internal waves in a rotating ocean. Okeanologia 18 (1978) 181-191.

[13] V. Varlamov, Y. Liu, Cauchy problem for the Ostrovsky equation. Discrete and Contin. Dyn. Systems 10 (2004) 731-751. https://doi.org/10.3934/dcds.2004.10.731

[14] V.E. Zakharov, Stability of periodic waves of finite amplitude on the surface of a deep fluid, Zh. Prikl. Mekh. Tekh. Fiz. 9 (1968), 86-94 (in Russian); J. Appl. Mech. Tech. Phys. 9 (1968), 190-194 (English translation).

Floating Offshore Energy Devices
Materials Research Proceedings 20 (2022) 26-30

Materials Research Forum LLC
https://doi.org/10.21741/9781644901731-4

A Review of Rapid Distortion Theory

Sudipta Lal Basu[1, a], Breiffni Fitzgerald[1,b] , Søren R.K.Nielsen[2,c] and Biswajit Basu[1,d*]

[1]Dept. of Civil, Structural & Environmental Engineering, Trinity College Dublin, Dublin, Ireland

[2]Department of Civil Engineering, Aalborg University, Denmark

[a]basus@tcd.ie, [b]Breiffni.Fitzgerald@tcd.ie, [c]srkn@civil.aau.dk, [d]basub@tcd.ie*

* corresponding author

Keywords: Rapid Distortion Theory, Turbulence, Wind Energy, Wind Farms

Abstract. With the depleting non-renewable fuel sources like coal and an ever-increasing demand for energy, we need to start looking into renewable energy sources. These are of paramount importance for a sustainable and green future. Wind Energy is one of the most important sources of renewable energy. But, setting up a wind farm requires considerable land area and land acquisitions are often faced with legal hurdles. This necessitates setting up offshore wind turbines. But, when we talk about offshore wind farms, we need to address the age-old phenomenon: "Turbulence". Presently, we are trying to develop enhanced controllers for wind farms which will increase the efficiency of the wind farms. The effects of rapidly changing wake aerodynamics i.e. breakdown of strong tip and hub vortices mixed up with low intensity turbulence in the inflow of the rotor and counter-rotation of the wake i.e. determinate velocity component in wake turbulence field will affect the overall performance of the wind farm. This paper provides a brief review on Rapid Distortion Theory (RDT) to model the turbulence.

Introduction

Batchelor, in European Turbulence Conference, 1986 at Lyon predicted that there could be no global theory for turbulence (other than that turbulent flows are governed by Navier-Stokes equations) because all turbulent flows are governed by the initial and boundary conditions. Batchelor & Proudman [1] had given a description of how turbulence is distorted when it passes rapidly through a region where large-scale straining motions are induced.

This review paper gives a brief description of the classification of the different types of problems of turbulence, a brief overview of the RDT and the errors associated with it. The application of the RDT to model the turbulence in case of offshore wind farms is also presented.

Classification of Turbulence

Hunt and Carruthers [4] in their paper "Rapid Distortion Theory and 'problems' of turbulence" have classified the turbulent flows based on their initial and boundary conditions.

Class I: Closed domains and deterministic boundary conditions
In this case, boundary surface $\mathscr{B}$ of the domain $\mathscr{D}$ consists of rigid stationary or moving surfaces. Turbulent motion could be induced due to the motion of the boundaries or due to the body forces. Example can be taken of a cylinder with a moving piston.

Floating Offshore Energy Devices   Materials Research Forum LLC
Materials Research Proceedings 20 (2022) 26-30   https://doi.org/10.21741/9781644901731-4

Class II: Open domains and statistical boundary conditions
In this class of turbulent flows, some, if not all of, the bounding surfaces of the domain $\mathscr{D}$ lie within the fluid itself. Let $\mathscr{E}$ denote the region outside domain $\mathscr{D}$. An example can be of the turbulence in the wake of the wind farm itself. This is further subdivided into:-

    i. No turbulence in $\mathscr{E}$: In this class the flow enters $\mathscr{D}$ with characteristic mean velocity U_0 and the turbulence could be generated due to instabilities if Reynolds Number is high.

    ii. Turbulence in $\mathscr{E}$ with significant mean flow from $\mathscr{E}$ to $\mathscr{D}$.

    iii. Turbulence in $\mathscr{E}$ without significant mean flow from $\mathscr{E}$ to $\mathscr{D}$

Class III: Initial conditions and changing boundary conditions
In both Class I and II above, the nature of turbulence is dependent on the boundary $\mathscr{B}$ of the domain $\mathscr{D}$ but with the assumption that the boundary conditions persist for long enough or they do not change that rapidly to cause an ongoing time-dependent change at $t = \tau$, τ being the instantaneous time at which we are interested in the nature of turbulence. This third classification considers the change in boundary conditions.

Mathematical developments of RDT
The linearization and the error analysis presented in this paper are from the works of Hunt and Carruthers [4]. The results of the same are studied and an effort will be made to implement the same in the case of wind farms. In the equations discussed, notations in bold indicate vector quantities.

Linearization
Reynolds suggested that the random velocity, pressure and vorticity fields $\mathbf{u}(\mathbf{x},t)$, $\mathbf{p}(\mathbf{x},t)$, $\boldsymbol{\omega}(\mathbf{x},t)$ as functions of position vector $\mathbf{x}$ and time, t can be divided into components of ensemble mean and fluctuating component i.e. $\mathbf{u} = \mathbf{U} + \mathbf{u'}$, $p = \rho(P + p')$, $\boldsymbol{\omega} = \boldsymbol{\Omega} + \boldsymbol{\omega'}$ where $\mathbf{U}$, P, $\boldsymbol{\Omega}$ are the ensemble means and $\mathbf{u'}$, p', $\boldsymbol{\omega'}$ are the fluctuating components. Discussion in this paper is strictly confined to incompressible flows in the absence of body forces. The governing equations for velocity and vorticity (in Einstein notation) are

$$\frac{\partial u'_i}{\partial t} + U_j \frac{\partial u'_i}{\partial x_i} + u'_j \frac{\partial U_i}{\partial x_j} = -\frac{1}{\rho}\frac{\partial p'}{\partial x_j} + \nu\nabla^2 u'_i - (NL)_{u'_i} \cdots \tag{1}$$

$$\frac{\partial \omega'_i}{\partial t} + U_j \frac{\partial \omega'_i}{\partial x_j} + u'_k \frac{\partial \Omega_i}{\partial x_k} - \omega'_k \frac{\partial U_i}{\partial x_n} - \Omega_n \frac{\partial u'_i}{\partial x_n} = \nu\nabla^2 \omega'_i + (NL)_{\omega'_i} \cdots \tag{2}$$

where, ν = kinematic viscosity, $\partial u'_i / \partial x_i = 0$, $\omega'_i = \varepsilon_{ijk}\,\partial u'_k / \partial x_j$ so that $\partial \omega'_k / \partial x_k = 0$
The nonlinear terms in the above equations are

$$(NL)_{u'i} = -\left[\frac{\partial(u'_k u'_i)}{\partial x_k} - \frac{\partial(\overline{u'_k u'_i})}{\partial x_k}\right] \cdots \tag{3}$$

Floating Offshore Energy Devices
Materials Research Proceedings **20** (2022) 26-30

Materials Research Forum LLC
https://doi.org/10.21741/9781644901731-4

$$(NL)_{\omega'i} = -u'_k \frac{\partial \omega'_i}{\partial x_k} + \omega'_j \frac{\partial u'_i}{\partial x_j} + \overline{\frac{u'_k \partial \omega'_i}{\partial x_j}} - \overline{\frac{\omega'_j \partial u'_i}{\partial x_j}} \cdots \qquad (4)$$

To linearize the governing equations, the nonlinear terms (NL) are to be ignored. An effort is then required to be made to find a solution for the linearized form of the equations by applying suitable boundary conditions. However, ignoring the (NL) terms will result in errors which in turn needs to be analysed so that solution of the linearized equation is acceptable and reflects the nearly acceptable physical scenario of the problem, if not exact.

While dealing with offshore wind-turbines, the mean velocity of the wind is significantly higher compared to its fluctuations. Apart from ignoring the (NL) terms, for air kinematic viscosity is of the order of 10^{-5} m^2/s and we can infer that viscosity will not have much effect on the final result, so it will be negligible and can be ignored. Thus, we can view this problem as part of Class-II(Type-ii) case of classification stated above.

Error Analysis
In order to estimate the error associated with the linearization, we begin with assuming a typical r.ms. velocity $u'_0 = \left(\frac{1}{3}\overline{u_i u_i}\right)^{1/2}$ and integral scale L_x for the large energy containing scales of turbulence and for small eddies with velocity scale u(l) and length scale l. U_0 and ΔU_0 are the typical values of the mean velocity and the change in mean velocity respectively over a typical length scale in $\mathscr{D}$. The two-point moment of velocity field, $R_{ij}(r) = \overline{u'_i(x)u'_j(x,r)}$ or the two-point structure function, $\Delta R_{ii} = \overline{(u_i(x) - u_i(x,r))^2}$ (r is the vector defining distance between the two points) are primarily calculated from the linearized equations. If ω' is used to calculate **u'** and R_{ii}, the conditions for linearization are different from using $\overline{\omega'_i}^2$ because of the requirement of specifying the scale of vorticity field contributing to the moment.

Batchelor [5], using Biot-Savart integral showed that $\Delta R_{ii}(x,r)$ can be expressed as an integral of $\overline{\omega'_k(r')\omega'_l(r'')}$, where r' and r'' are the displacement vectors corresponding to k and l respectively. For high Reynolds number it can be expressed in terms of rate of dissipation per unit mass, ε by the following relation

$$\Delta R_{ii}(x,r) \sim \int_0^l \varepsilon^{2/3} \hat{r}^{-4/3} \hat{r} \, d\hat{r} \sim \varepsilon^{2/3} l^{2/3} \cdots \qquad (5)$$

where $\hat{r} = |(r' - r'')|$, $|r| = l$, $|r' - r''| < L_x$, $\overline{\omega'_k(r')\omega'_l(r'')} \sim \varepsilon^{2/3} \hat{r}^{-4/3}$

It can be seen from eq. (5) that although the co-relation of vorticities is significant at small separations of $\hat{r}$, the contribution from smaller-scale vorticity is comparable to that from length-scales of order l. Thus, the contribution to the eddies could come from either vortex sheets separated by L_x or from smooth distributions of vorticities on a scale l or from both.

The effect of non-linear terms (4) are to be estimated over the appropriate length scale (l) of the velocity field and the time period (T$_\mathscr{D}$) over which the distortion is applied. The second term of eq. (4) related to the stretching of the fluctuating vorticity by the fluctuating velocity is of the order $\varepsilon^{1/3} l^{-2/3}$. Over the time period T$_\mathscr{D}$, the relative change in the linear and non-linear terms of the vorticity w.r.t. the initial vorticity ω_0, is given by $\Delta\omega_{Lin}/\omega_0 \sim (\Delta U/L_\mathscr{D})T_\mathscr{D}$, where ΔU is the change in mean velocity and $\Delta\omega_{NL}/\omega_0 \sim (u'(l)/l)T_\mathscr{D}$, where $L_\mathscr{D}$ is the length-scale of domain $\mathscr{D}$. The criterion for ignoring the nonlinear vortex stretching term is therefore given by the following equation.

$$\frac{u'(l)}{l} \sim \varepsilon^{\frac{1}{3}} l^{-\frac{2}{3}} \ll max\left(\frac{\Delta U}{L_{\mathscr{D}}}, \frac{1}{T_{\mathscr{D}}}\right) \dots \tag{6}$$

In terms of the characteristic velocity of the energy containing eddies u'_0, the inertial range scaling $u(l) \sim u'_0 (l/L_x)^{1/3}$, eq. (6) can be reframed as

$$\frac{u'_0}{L_x}\left(\frac{l}{L_x}\right)^{-2/3} \ll max\left(\frac{\Delta U}{L_{\mathscr{D}}}, \frac{1}{T_{\mathscr{D}}}\right) \tag{7}$$

The total strain defined by $\beta = T_{\mathscr{D}} \Delta U / L_{\mathscr{D}}$ and the relative rate of strain defined by

$$\mathscr{S}^* = (\Delta U / L_{\mathscr{D}}) T_L, where \ T_L = \frac{L_x}{u'_0} \dots \tag{8}$$

are the two dimensionless quantities characterizing the energy containing eddies. It can be concluded that for weaker strain rates $\mathscr{S}^* \leq 1$,

$$(\beta / \mathscr{S}^*) \ll 1, or \ T_{\mathscr{D}} \ll T_L \tag{9}$$

In case the strain rate is strong, $\mathscr{S}^* \geq 1$, then

$$(\beta / \mathscr{S}^*) \gg 1 \dots \tag{10}$$

implying $T_{\mathscr{D}}/T_L$ is arbitrary. Satisfaction of eq. (7) implies that the effects of random straining with large time scales tend to be negligible. Thus, this condition is the essential criterion for the RDT to be valid in case of rapidly changing turbulent flows. Thus, the linearization can be justified only if the strain-rate is significantly large or the period of distortion is significantly short.

The anisotropy caused by the mean strain can be under-estimated or over-estimated by linearization. Therefore, the criterion of eq. (7) which enables us to neglect the non-linear terms is applicable to velocity and vorticity only if modified to take into account the reduce straining in some direction and the nonlinear rotation. This modifies the equation to

$$\frac{u_0}{L_x}\left(\frac{l}{L_x}\right)^{-2/3} \ll max\left(\frac{\Delta U}{L_{\mathscr{D}}}\theta(T_{\mathscr{D}}), \frac{1}{T_{\mathscr{D}}}\right) \dots \tag{11}$$

where, $\theta(T_{\mathscr{D}}) = \exp\left((\lambda_{max} - \lambda_{min})T_{\mathscr{D}}\right)$ where λ_{max} & λ_{min} are the moduli of the maximum and minimum values of principal strain of the mean flow field $\partial U_i / \partial x_j$. Thus, eq. (10) condition for changes to $\mathscr{S}^* \theta \geq 1$. Batchelor [2] proved that for strong enough compressive strains, $\partial U_i / \partial x_j \ \alpha \ \delta_{ij}$ and $\theta = 1$, and if the criteria is satisfied, the non-linear terms can be neglected for all time. However, if for any non-isotropic strain the value of θ increases with time, the non-linear term can no more be neglected.

Summary
In the present project, we are focusing on the application of the RDT to model the turbulence for the wind farms. The linearization of Navier-Stokes equation using RDT and the errors associated are to be analysed in greater details in this context. Simply ignoring the non-linear terms is of no good if it leads to a great level of approximation errors. Choosing the relevant length scale and time scales of the vortices will play an important role in the approximation using RDT.

Floating Offshore Energy Devices
Materials Research Proceedings **20** (2022) 26-30

Materials Research Forum LLC
https://doi.org/10.21741/9781644901731-4

References

[1] G.K. Batchelor, I. Proudman, The effects of Rapid Distortion of a fluid in turbulent motion, Q. J.Mech. Appl. Maths, 7 (1954), 83-103. https://doi.org/10.1093/qjmam/7.1.83

[2] G.K. Batchelor, The effective pressure exerted by a gas in turbulent motion, In Vistas in Astronomy (ed. A.Beer), 1(1955), 290-295. https://doi.org/10.1016/0083-6656(55)90038-6

[3] G.K. Batchelor, I. Proudman, The large-scale structure of homogeneous turbulence, Phil. Trans. R. Soc. Lond.,A 248(1956), 369-405. https://doi.org/10.1098/rsta.1956.0002

[4] J.C.R. Hunt, D.J. Carruthers, Rapid Distortion Theory and the 'problems' of turbulence, J. Fluid Mech. 212 (1990) 497-532. https://doi.org/10.1017/S0022112090002075

[5] G.K. Batchelor, An Introduction to Fluid Dynamics, Cambridge University Press, 1967

[6] A.S.Monin, A.M.Yaglom, Statistical Theory of Turbulence, Vol. I, MIT Press, 1971.

[7] A.S. Monin, A.M. Yaglom, Statistical Theory of Turbulence, Vol. II, MIT Press, 1975.

Floating Offshore Energy Devices
Materials Research Proceedings **20** (2022) 31-38

Materials Research Forum LLC
https://doi.org/10.21741/9781644901731-5

On the Importance of High–Fidelity Numerical Modelling of Ocean Wave Energy Converters under Controlled Conditions

Christian Windt[1*], Josh Davidson[2], Nicolas Faedo[1], Markel Penalba[3] and John V. Ringwood[1]

[1]Centre for Ocean Energy Research, Maynooth University North Campus, Maynooth, Co. Kildare, Ireland

[2]Department of Fluid Mechanics, Faculty of Mechanical Engineering Budapest University of Technology and Economics, Budapest, Hungary

[3]Department of Mechanical Engineering and Industrial Production Mondragon Unibertsitatea, Mondragon, Pais Vasco, Spain

*e-mail: christian.windt.2017@mumail.ie

Keywords: Wave Energy Converter, Numerical Modelling, BEM, CFD, Control

Abstract. Numerical modelling tools are commonly applied during the development and optimisation of ocean wave energy converters (WECs). Models are available for the hydrodynamic wave structure interaction, as well as the WEC sub–systems, such as the power take–off (PTO) model. Based on the implemented equations, different levels of fidelity are available for the numerical models. Specifically under controlled conditions, with enhance WEC motion, it is assumed that non-linearities are more prominent, re- quiring the use of high–fidelity modelling tools. Based on two different test cases for two different WECs, this paper highlights the importance of high–fidelity numerical modelling of WECs under controlled conditions.

Introduction

The growing recognition of human induced global warming has fuelled the research and development (R&D) of novel technologies to harness renewable energy resources. Amongst these resources, marine renewable energy, and specifically ocean wave energy, shows significant potential to contribute to the global energy supply [1]. To increase the efficiency and, thereby, the economical feasibility of WECs, devices should be equipped with energy maximising control systems (EMCSs) [2]. Since the objective of EMCSs is to drive the system towards resonance with the incoming wave field, WEC motion of a controlled device is enhanced (see Figure 1), and the power conversion is increased.

During the design and optimisation of WECs, researchers and engineers rely on physical wave tank (PWT), as well as numerical wave tank (NWT) tests. Generally, by testing in a real physical environment, PWTs allow all the relevant details of the wave-structure interaction (WSI) to be captured. However, although still cheaper compared to open ocean trials, PWT experiments are associated with higher costs compared to NWT experiments [3]. The main cost drivers for PWT tests are instrumentation, construction of the prototype, test facilities, and staff. Additionally, the accuracy of PWT experiments potentially suffer from peculiarities of the test facility, such as reflections from the tank walls, friction in mechanical device components, measurement noise, and scaling effects.

Floating Offshore Energy Devices   Materials Research Forum LLC
Materials Research Proceedings **20** (2022) 31-38   https://doi.org/10.21741/9781644901731-5

Overcoming the drawbacks of high costs, measurement noise, mechanical friction, and, to a great extent, scaling effects, NWTs provide powerful tools for the analysis of WECs. Depending on the implemented equations for the solution of the WSI problem, different levels of fidelity, at different levels of computational cost, can be achieved [4]. Hydrodynamic, lower-fidelity models, such as Boundary Element Method (BEM) - based NWTs, neglecting non-linear effects, such as viscosity, are associated with minimal computational cost, and are valuable tools for parametric studies or exhaustive-search optimisation algorithms. However, due to the required linearisation of the hydrodynamic equations, lower-fidelity models are only valid when considering linear conditions, i.e. small amplitude waves and device motions. Contrary,

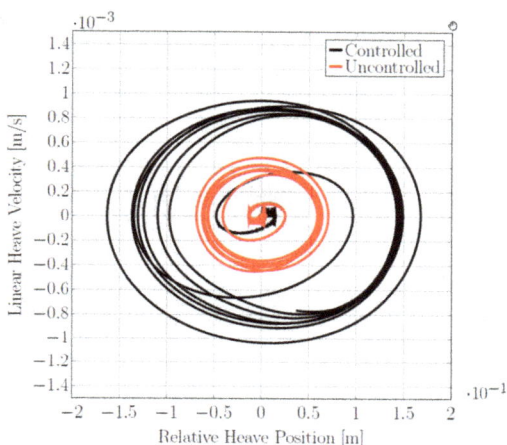

Figure 1: Operation space of uncontrolled and controlled WEC devices under regular wave excitation.

higher-fidelity NWTs, such as Computational Fluid Dynamics (CFD)-based numerical wave tanks (CNWTs), are able to capture all relevant hydrodynamic non-linearities by numerically solving the Navier-Stokes equations. Thus, CNWTs are valid over a wider range of test conditions, compared to lower-fidelity models.

Equally, when considering the sub-systems of a WEC device, such as the power–take off (PTO) or the mooring system, a range of numerical models with varying degree of fidelity are available and can be coupled with the hydrodynamic model [5]. Generally, the implementation of high–fidelity models of the WEC sub–systems is desired when employing CNWTs, to prevent the lower–fidelity sub–system models from undermining the accuracy of the high–fidelity hydrodynamic model.

The importance of high–fidelity modelling of the WSI, as well as the WEC sub–system, specifically under controlled conditions, will be investigated in the present paper. To that end, two different case studies are considered, analysing two different WECs: (1) the Wavestar device; (2) a generic heaving point absorber (HPA) type WEC (see Figure 2). In the first case study, considering the Wavestar device, the influence of different design and evaluation frameworks for EMCSs will be investigated, employing three different EMCSs of varying *aggressiveness*. In the second case study, considering the HPA-type WEC, the influence of the fidelity of the hydrodynamic model and the coupled PTO model is investigated.

The remainder of the paper is organised as follows: Section 2 details the low– and high– fidelity numerical wave tanks employed in the two case studies. Furthermore Section 2 presents a description of the employed PTO models. The results of the first case study, assessing the evaluation framework of EMCSs, will be presented and discussed in Section 3.

The second case study, investigating the influence of the fidelity of the hydrodynamic and PTO model, is discussed in Section 4. Finally, conclusions are drawn in Section 5.

Floating Offshore Energy Devices   Materials Research Forum LLC
Materials Research Proceedings 20 (2022) 31-38   https://doi.org/10.21741/9781644901731-5

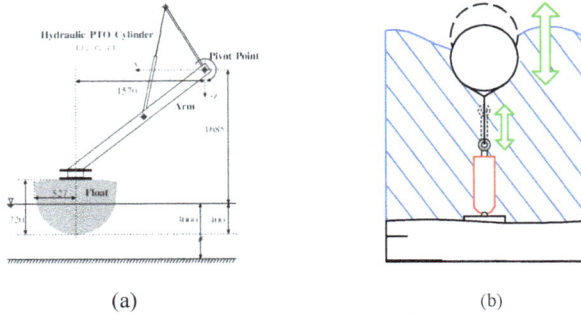

(a)   (b)

Figure 2: The considered WECs: (a) the Wavestar device; (b) a generic heaving point absorber (HPA) type WEC

Numerical of WECs

BEM-based NWTs (BNWTs). Linear time-domain hydrodynamic models based on the BEM use, in general, Cummins equation [6], following:

$$M\ddot{x}(t) = -s_h x + \mathcal{F}_{exc}(t) - \mu_\infty \ddot{x}(t) - \int_{-\infty}^{\infty} K_{rad}(t-\tau)\dot{x}(\tau)d\tau + u(t), \tag{1}$$

where x, $\dot{x}$ and $\ddot{x}$ are the displacement, velocity and acceleration of the WEC, respectively. M is the mass of the WEC, s_h the hydrostatic stiffness, $\mathcal{F}_{exc}(t)$ the excitation force, μ_∞ the added-mass at infinite frequency, $\zeta(t)$ the radiation impulse response function (IRF), and $u(t)$ the control law (PTO force).

The linear hydrodynamic model can be extended to include non–linear effects, such as nonlinear Froude–Krylov (FK) forces or viscous effects. Non–linear FK forces can be included using, e.g. the computationally efficient algebraic solution as presented in [7]. Viscous effects can be incorporated by using a Morison-like equation [8],

$$F_{visc} = -\frac{1}{2}\rho C_d A_d(t)|\dot{z}_d - \dot{\eta}|(\dot{z}_d - \dot{\eta}), \tag{2}$$

where ρ is the density of water, C_d the drag coefficient, A_d the instantaneous cross-sectional area of the device, and $\dot{\eta}$ the velocity of the undisturbed water particles.

CNWT. The CNWT simulations are performed using the open source CFD toolbox Open-FOAM. In OpenFOAM, the incompressible Reynolds Averaged Navier-Stokes (RANS) equations (3) and (4) are solved using the finite volume method.

$$\nabla \cdot \rho U = 0 \tag{3}$$

$$\frac{\partial(\rho U)}{\partial t} + \nabla \cdot (\rho U U) = -\nabla p + \nabla \cdot T + \rho f_b \tag{4}$$

Equation (3) is the continuity equation, describing the conservation of mass, and equation (4) is the momentum equation. describing the conservation of momentum. In equations (3) and (4), U

Floating Offshore Energy Devices
Materials Research Proceedings **20** (2022) 31-38

Materials Research Forum LLC
https://doi.org/10.21741/9781644901731-5

denotes the fluid velocity, p the fluid pressure, ρ the fluid density, T the stress tensor, and f_b external forces such as gravity or PTO forces.

To account for the two phase flow, the volume of fluid method, proposed in [9], is used, following:

$$\frac{\partial \alpha}{\partial t} + \nabla \cdot (u\alpha) + \nabla \cdot [u_r \alpha (1 - \alpha)] = 0 \tag{5}$$

$$\Phi = \alpha \Phi_{\text{water}} + (1 - \alpha)\Phi_{\text{air}} , \tag{6}$$

where α denotes the volume fraction of water, $u_r(t)$ is the relative velocity between the liquid and gaseous phases [10], and Φ is a specific fluid quantity, such as density. The free surface elevation is monitored by extracting the iso-surface of the volume fraction $\alpha = 0.5$.

PTO. As for the hydrodynamic model, different levels of fidelity are available for the modelling of the PTO system of a WEC. One of the simplest models describes the PTO as a spring damper system, following:

$$F_{\text{PTO}}(t) = Kx(t) + B\dot{x}(t) \tag{7}$$

where B is a damping coefficient and $\dot{x}(t)$ the linear velocity of the hydraulic PTO cylinder, K is a spring stiffness, and $x(t)$ the linear motion of the PTO. The damping and stiffness coefficient either represent the mechanical characteristics of, say, a hydraulic cylinder, or B and K are representing the EMCSs and are optimised for maximum energy absorption. Higher–fidelity PTO models are available, including e.g. a hydraulic transmission system and an electrical generator [11]. The mathematical model for the hydraulic cylinder may include end-stop constraints, friction losses, and compressibility and inertia effects, providing the final PTO force, following:

$$F_{PTO} = A_p \Delta p + F_{fric} + F_l \tag{8}$$

where A_p is the piston area, Δp the pressure difference between the different cylinder chambers, F_{fric} the friction force and F_l the inertia force. For a detailed description of the individual effects, influencing the PTO force, the interested reader is referred to [11].

Case study 1: Assessment of the evaluation framework for EMCSs

In classical control applications, the mathematical models, used for the controller design, are often linearised around a desired operational point, according to the process under analysis. The controller is subsequently synthesised to drive the system towards this point and, thus, in the neighbourhood of this operational point, the linearising assumption is obeyed. The large amplitude motions, induced by a reactive WEC controller, may result in viscous drag and other non-linear hydrodynamic effects. Thus, contrary to the aforementioned classical control applications, the energy-maximising operating conditions do not comply with the linear assumption in the control design model.

This contradiction between the control objective and the underlying mathematical model raises the question if the common practice of designing a controller in a linear design environment can deliver optimal reactive controllers for the application in physical, non-linear operational conditions.

In this case study, a CNWT and linear BNWT model[1], as described in Section 2, are employed to investigate the influence of different numerical evaluation frameworks on the performance

---

[1] Note that, for this case study, any non-linearities, such as non-linear FK forces or viscous drag effects, are neglected.

evaluation of EMCSs for the Wavestar device (see Figure 2a). The performance of the EMCSs will be evaluated by comparing the dynamics of the WEC subject to three different EMCSs: (1) moment-based energy-maximising control [12]; (2) reactive output feedback control; (3) resistive output feedback control.

The three EMCSs will drive the WEC away from the linear assumption in the hydrodynamic model, dependent on the aggressiveness of the controller, with the resistive controller being the least aggressive and the moment-based control the most aggressive.

EMCSs. The main objective of a wave energy device is to harvest energy from the incoming wave field. Therefore, the optimal control objective is to maximise the absorbed energy over a time interval [t, t + T], while respecting the physical limitations of the device/PTO on excursion $x(t)$, velocity $\dot{x}(t)$, and PTO force $u(t)$. Consequently, the optimal control objective can be formulated as

$$u^{\max}(t) = \arg\max_{u(t)} \int_t^{t+T} u(\tau)\dot{x}(\tau)d\tau \tag{9}$$

The optimal control is a moment-based WEC formulation [12] which allows an efficient computation of the optimal control law $u^{\max}$ in real-time based on the solution of the following inequality constrained quadratic program:

$$L_u^{\max} = \arg\max_{L_u} \ -\frac{1}{2}L_u\Phi_\varphi^\mathcal{R}L_u^\mathsf{T} + \frac{1}{2}L_{exc}\Phi_\varphi^\mathcal{R}L_u^\mathsf{T}, \tag{10}$$

The reader is referred to [12] for the formal definition (and corresponding proofs) of the matrices involved in the QP problem of (10).

Additionally to the moment-based controller, less aggressive EMCSs, i.e. reactive and resistive controllers, are considered herein. For the reactive control case, the PTO force follows

$$u(t) = K_{\text{opt}}x(t) + B_{\text{opt}}\dot{x}(t) \tag{11}$$

where B_{opt} is the optimal damping coefficient and K_{opt} is the optimal spring stiffness. The optimal PTO coefficients have been determined through exhaustive search optimisation using the BNWT model. For the resistive control case, only the second term, $B_{\text{opt}}\dot{x}(t)$, is considered. As in the case of reactive control, the optimal damping coefficient has been determined through exhaustive search optimisation using the BNWT model.

Results. For each EMCSs, simulations were performed in the BNWT and CNWT, resulting in a total of six simulations. Extracting the PTO displacement, velocity, and the PTO force, the normalised root mean squared deviation (nRMSD), as defined in equation (12), can be calculated. y_{BNWT} denotes the results from the BNWT experiment, y_{CNWT} results from the CNWT experiment, and N is the number of samples.

$$\text{nRMSD} = \sqrt{\frac{\Sigma(y_{\text{CNWT}}-y_{\text{BNWT}})^2}{N}} \cdot \frac{1}{\max(y_{\text{BNWT}})-\min(y_{\text{BNWT}})} \cdot 100\% \tag{12}$$

For a qualitative comparison between the three different EMCSs, Figure 3 shows the PTO displacement, extracted from the BNWT experiment. In Figure 3, a clear trend can be observed.

Floating Offshore Energy Devices
Materials Research Proceedings **20** (2022) 31-38

Materials Research Forum LLC
https://doi.org/10.21741/9781644901731-5

The moment-based and the reactive controller controller increases the PTO displacement, compared to the resistive controller.

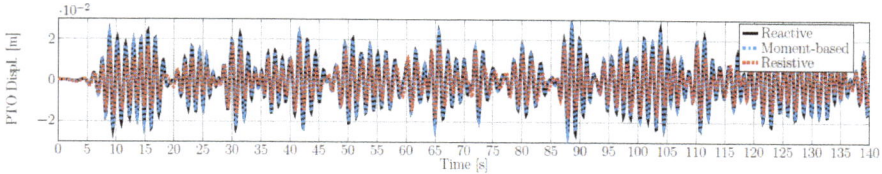

Figure 3: Time traces of the PTO displacement x(t) from the BNWT experiment

The resulting nRMSD for the PTO displacement, velocity, and force are listed in Table 1. Overall smallest values of the nRMSD can be observed for the case of a resistively controlled device. For this control strategy, smallest deviations are indeed expected, since the device effectively acts as a wave follower, reducing the influence of non-linear hydrodynamic effects, such as non-linear Froude-Krylov forces or viscous drag effects.

For the cases with reactive and moment-based control, the values for the nRMSD in all PTO quantities increase. In Table 1, it can be seen that the differences between the PTO quantities for the two different controllers are relatively small. This can be attributed to the fact that both state (displacement) and input (PTO force) constraints are inactive for this particular input wave, i.e. both controllers effectively reflect the unconstrained optimal energy-absorption conditions. That said, it is expected that also the differences between the BNWT and CNWT are similar.

Table 1: nRMSD between the results for PTO cylinder displacement, velocity and PTO force from the BNWT and CNWT for the different EMCSs

nRMSD [%]		Moment-based	Reactive	Resistive
PTO cylinder displacement	x	4.53	4.50	3.20
PTO cylinder velocity	$\dot{x}$	3.79	3.76	2.47
PTO cylinder Force	$u(t)$	4.33	4.30	2.47

Case study 2: Coupled numerical wave tank and PTO models
Based on a generic, reactively controlled, HPA-type WEC (see Figure 2b), this case study will investigate the influence of different levels of model fidelity of the hydrodynamic, as well as the PTO model. To that end, four different numerical models are considered: (1) a CNWT coupled with a high–fidelity wave-to-wire (W2W) model (CNWT+W2W); (2) a CNWT coupled with an idealised, linear spring–damper PTO model (CNWT+iPTO); (3) a linear BNWT coupled with an idealised, linear spring–damper PTO model (BNWT+iPTO); (4) a non-linear BNWT with a drag coefficient of $C_d = 1$ coupled with an idealised, linear spring–damper PTO model (nlBNWT$_{C_d=1}$+W2W).

Results. The time–average absorbed and generated power values from the numerical models (2)–(4) are compared against the values from the CNWT+W2W. Absorbed power refers to the mechanical power directly absorbed from ocean waves, while generated power refers to the final electric power output. Since the CNWT+W2W accounts for all relevant hydrodynamic non-linearities, as well as the occurring non-linearities in the PTO drivetrain, this model is considered as the benchmark in this case study.

For a quantitative comparison, Table 2 shows that the relative deviation, ε , between the CNWT+W2W model and models (2)–(4) in absorbed power. Considering only the time– average, absorber power, relatively small differences ($\varepsilon < 8\%$) can be observed between models (1), (2), and (4), indicting a sufficient accuracy of the non–linear BNWT with a drag coefficient of $C_d = 1$. Larger deviations can be observed between the linear BNWT, model (3), and the benchmark, model (1). This indicates that a linear hydrodynamic model is not able to capture all relevant hydrodynamic effects.

Considering the generated power for the comparison between the numerical models, the influence of the non–linear W2W model becomes visible. For the two numerical models including an idealised, linear, spring-damper type PTO model (i.e. model (2) and (3)), relative differences to the benchmark model (1) of up to 97.52% are calculated. For the two W2W models only a difference of $\varepsilon < 3\%$ can be observed, which is consistent with the findings for the absorbed power.

Table 2: Time-averaged absorbed and generated power obtained from the CNWT+W2W model and the percentage difference (ε) to the other considered models.

Model			Reactive control P_{abs}^{av}	P_{gen}^{av}
(1)	CNWT+W2W	[kW]	23.76	13.99
(2)	CNWT+iPTO	[ε (%)]	3.03	78.36
(3)	BNWT+iPTO	[ε (%)]	16.32	97.52
(4)	nlBNWT$_{C_d=1}$+W2W	[ε (%)]	-7.60	-2.87

Conclusions
In this paper, two different case studies are presented, highlighting the importance of (consistent) high-fidelity modelling of WECs under controlled conditions. From the presented results, two main conclusions can be drawn:
1. Considering aggressive EMCSs for WEC control, driving the system further away from the assumptions in the linear hydrodynamic model, high–fidelity hydrodynamic models have to be employed for the assessment of the performance of the EMCSs. Omitting high–fidelity hydrodynamic modelling in the evaluation stage of the control design can lead to an over–prediction of the WEC performance.
2. The holistic performance of a WEC can only be evaluated in high–fidelity by means of a comprehensive W2W simulation platform, where both high–fidelity hydrodynamic and PTO models are coupled. Minor inaccuracies in either of these major stages of the W2W model can result in significant inaccuracy in generated power estimation.

Acknowledgment
This material is based upon works supported by the Science Foundation Ireland under Grant No. 13/IA/1886. The research reported in this paper was also supported by the Higher Education Excellence Program of the Ministry of Human Capacities in the frame of Water science & Disaster Prevention research area of Budapest University of Technology and Economics (BME FIKP-VÍZ)

References

[1] A. F. de O. Falcão, "Wave energy utilization: A review of the technologies," *Renewable and Sustainable Energy Reviews*, vol. 14, pp. 899–918, 2010. https://doi.org/10.1016/j.rser.2009.11.003

[2] J. V. Ringwood, G. Bacelli, and F. Fusco, "Energy-maximizing control of wave- energy converters: The development of control system technology to optimize their operation," *IEEE Control Systems*, vol. 34, pp. 30–55, 2014. https://doi.org/10.1109/MCS.2014.2333253

[3] J. W. Kim, H. Jang, A. Baquet, J. O'Sullivan, S. Lee, B. Kim, and H. Jasak, "Technical and economic readiness review of CFD-based numerical wave basin for offshore floater design," in *Proceedings of the 2016 Offshore Technology Conference, Houston, TX, USA*, 2016. https://doi.org/10.4043/27294-MS

[4] M. Penalba, G. Giorgi, and J. V. Ringwood, "Mathematical modelling of wave energy converters: a review of nonlinear approaches," *Renewable and Sustainable Energy Reviews*, vol. 78, pp. 1188–1207, 2017. https://doi.org/10.1016/j.rser.2016.11.137

[5] C. Windt, J. Davidson, and J. Ringwood, "High-fidelity numerical modelling of ocean wave energy systems: A review of computational fluid dynamics-based numerical wave tanks," *Renewable and Sustainable Energy Reviews*, vol. 93, pp. 610 – 630, 2018. https://doi.org/10.1016/j.rser.2018.05.020

[6] W. E. Cummins, "The impulse response function and ship motions," DTIC Document, Tech. Rep., 1962.

[7] G. Giorgi and J. V. Ringwood, "Computationally efficient nonlinear Froude–Krylov force calculations for heaving axisymmetric wave energy point absorbers," *Journal of Ocean Engineering and Marine Energy*, vol. 3, no. 1, pp. 21–33, February 2017. https://doi.org/10.1007/s40722-016-0066-2

[8] J. R. Morison, M. P. O'Brien, J. W. Johnson, and S. A. Schaaf, "The forces exerted by surface waves on piles," *Petroleum Trans., AIME. Vol. 189, pp. 149-157*, 1950. https://doi.org/10.2118/950149-G

[9] C. W. Hirt and B. D. Nichols, "Volume of Fluid (VOF) Method for the Dynamics of Free Boundaries," *Journal of Computational Physics*, vol. 39, pp. 201–225, 1981. https://doi.org/10.1016/0021-9991(81)90145-5

[10] E. Berberović, N. P. van Hinsberg, S. Jakirlić, I. V. Roisman, and C. Tropea, "Drop impact onto a liquid layer of finite thickness: Dynamics of the cavity evolution," *Physical Review E*, vol. 79, pp. 036306–1 – 036306–15, 2009. https://doi.org/10.1103/PhysRevE.79.036306

[11] M. Penalba, J. Davidson, C. Windt, and J. V. Ringwood, "A high-fidelity wave-to- wire simulation platform for wave energy converters: Coupled numerical wave tank and power take-off models," *Applied energy*, vol. 226, pp. 655–669, 2018. https://doi.org/10.1016/j.apenergy.2018.06.008

[12] N. Faedo, G. Scarciotti, A. Astolfi, and J. V. Ringwood, "Energy-maximising control of wave energy converters using a moment-domain representation," *Control Engineering Practice*, vol. 81, pp. 85–96, 2018. https://doi.org/10.1016/j.conengprac.2018.08.010

Floating Offshore Energy Devices
Materials Research Proceedings **20** (2022) 39-46

Materials Research Forum LLC
https://doi.org/10.21741/9781644901731-6

Consistent Modal Calibration of
a Pendulum-Type Vibration Absorber

Jan Høgsberg

Department of Mechanical Engineering, Technical University of Denmark, Nils Koppels Alle, building 404, DK 2800 Kongens Lyngby, Denmark

jhg@mek.dtu.dk

Keywords: Vibration Absorber, Structural Dynamics, Modal Analysis, Absorber Calibration

Abstract. Pendulum absorbers are installation in offshore wind turbines to mitigate excessive vibration amplitudes from wind and wave loading. The pendulum damper is placed inside the tower and attached to the structure at two distinct points: The tower top, where the pendulum arm is fixated, and at the position of the pendulum mass, which is connected to the tower wall by the damper. The present paper derives a modal calibration principle, which consistently accounts for different points of attachment for the absorber stiffness and damping.

Introduction

Offshore wind turbines are among the most popular and effective renewable energy sources available today. In local areas with large water depth, such as in Norway, the classic foundation types are infeasible and thus floating platforms seem to be the most viable alternative. Absorber devices, such as a pendulum-type vibration absorber, may therefore be installed to compensate for the increased flexibility. The damping of offshore wind turbines is considered in greater detail in references [1-4].

Vibration absorbers are calibrated with respect to a single targeted vibration mode, with a well-defined natural frequency and mode shape [1, 5-7]. A modal system reduction then results in a two-degree-of-freedom (2-dof) model with a single mode coupled with the single-mass absorber. For the pendulum absorber the apparent stiffness connects the absorber mass to the tower-top, while the dashpot transfers the absorber force to the tower wall. This non-collocation implies a modelling error in the modal reduction, that may be taken into account be redefining the 'undamped' structure as the compound system with the absorber dashpot fully locked. When using the mode shape for this augmented undamped structural model, the scalar structural equation becomes less sensitive to any feedback from other vibration modes [8]. Furthermore, it retains the dynamic model associated with vanishing absorber damping as a case that can be used to calibrate for residual mode correction coefficients, which include the modal interaction with other vibration modes [9]. This correction is however not considered in the present analysis, which simply truncates any modal series representation to its single dominant term. The proposed calibration formulae in the present paper can be used in practice to improve the performance of pendulum dampers, as well as more advanced absorbers with flexible appendages.

The structural model

Assume a build-up FE model of the offshore wind turbine (owt) structure in Fig.1(a), with the pendulum-type vibration absorber attached inside the tower. When the pendulum behavior is linearized, the equation of motion can be written as

Floating Offshore Energy Devices
Materials Research Proceedings 20 (2022) 39-46

Materials Research Forum LLC
https://doi.org/10.21741/9781644901731-6

$$\mathbf{M\ddot{q} + C\dot{q} + Kq = f} \tag{1}$$

with $\mathbf{M}$, $\mathbf{C}$ and $\mathbf{K}$ representing the model mass matrix, damping matrix and stiffness matrix, respectively, while the vector $\mathbf{q}$ contains the general dofs of the combined structure-absorber dynamic model.

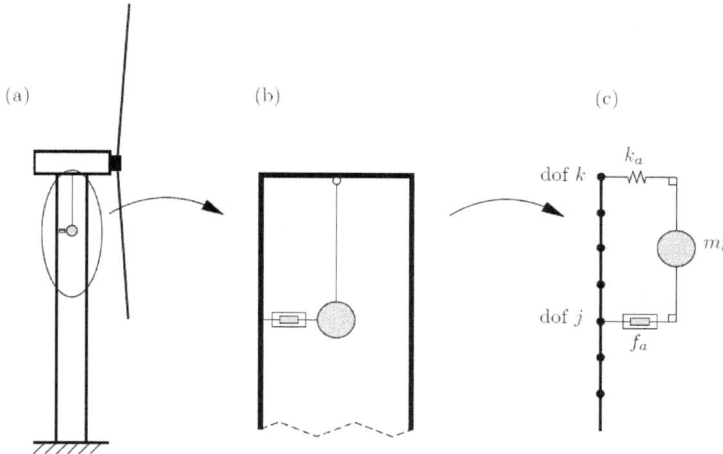

Figure 1: (a) Wind turbine with tower pendulum absorber, (b) close-up of pendulum with damper and (c) equivalent linearized model with damper force f_a at dof j and apparent spring from pendulum arm at an upper dof k.

The system is split by dividing the dofs into those associated with the structure and the supplemental dofs representing the absorber. It should be noted that common dofs are included in the structure. Hereby, the displacement vector can be decomposed as

$$\mathbf{q} = \begin{bmatrix} \mathbf{u}_s \\ u_a \end{bmatrix} \tag{2}$$

whereby the system matrices are correspondingly split as follows,

$$\mathbf{M} = \begin{bmatrix} \mathbf{M}_s & \mathbf{0} \\ \mathbf{0}^T & m_a \end{bmatrix} \quad , \quad \mathbf{C} = \begin{bmatrix} c_a \mathbf{d}\mathbf{d}^T & -c_a \mathbf{d} \\ -c_a \mathbf{d}^T & c_a \end{bmatrix} \quad , \quad \mathbf{K} = \begin{bmatrix} \mathbf{K}_s + k_a \mathbf{b}\mathbf{b}^T & -k_a \mathbf{b} \\ -k_a \mathbf{b}^T & k_a \end{bmatrix} \tag{3}$$

with subscripts s and a referring to *structure* and *absorber*, respectively. The non-identical connectivity vectors $\mathbf{b}$ and $\mathbf{d}$ are zero vectors with a single unit value at dof k and j, respectively.

Classic tuning method

The common tuning principle for vibration absorbers follows from the analysis of the so-called tuned mass damper (TMD) [5, 6], which corresponds to the pendulum damper in Fig. 1(c) with $j=k$. The dynamics of the structure are represented by the vibration mode $\mathbf{\bar{u}}_0$ governed by

$$\left(\mathbf{K}_s - \omega_0^2 \mathbf{K}_s\right)\bar{\mathbf{u}}_0 = 0 \tag{4}$$

where ω_0 is the corresponding natural frequency of the virgin host structure. The calibration formulae for the classic TMD are

$$\kappa_0 = \frac{\mu_0}{(1+\mu_0)^2} \quad , \quad \beta_0 = \sqrt{2\mu_0^3(1-\mu_0)} \tag{5}$$

in which the modal mass ratio $\mu_0 = m_a/m_0$ determines the absorber mass m_a relative to the modal mass m_0, the corresponding modal stiffness ratio $\kappa_0 = k_a/k_0$ defines the absorber stiffness k_a relative to the modal stiffness k_0, while the damper ratio $\beta_0 = c_a/\sqrt{k_0 m_0}$ represents the absorber viscous coefficient relative to the geometric mean of the modal mass and stiffness.

The modal mass m_0 and stiffness k_0 are not uniquely defined, as the absorber mass is attached at either dof j or k via the dashpot or spring, respectively. Figure 2 shows the frequency response curves for the top tower deflection u_{top} in (a) and the absolute absorber motion u_a in (b). The mode shape is normalized to unity at the top dof k for the red curve in Fig. 2, while the magenta curve represents the TMD tuning when the mode shape is normalized relative to the lower dof j, at which the absorber damper is connected.

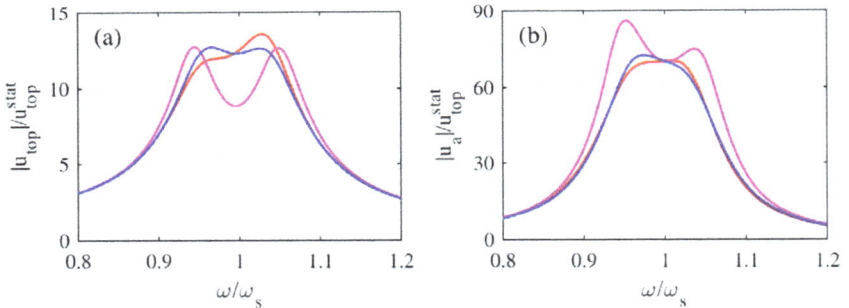

Figure 2: (a) Frequency amplitude curve for top tower motion (a) and for pendulum mass deflection (b).

Since the absorber mass m_a is constant, the modal mass ratio is $\mu_0 = 0.0154$ (red curve) for a top tower attachment at dof k, while at dof j it reduces to almost half $\mu_0 = 0.008$ (magenta curve) because of the reduced deflection at the lower position. It is seen from Fig. 2 that the assumed top tower attachment (red curves) leads to an inclined response frequency curve in (a) with clear peak and an almost flat and minimized pendulum deflection amplitude in (b). With respect to the lower attachment dof j (magenta curves) the response curve in (a) is almost equally balanced but has clearly too little damping, while its pendulum amplitude curve in (b) has a significant overshoot. The blue curves represent the proposed calibration described next, which seems to provide the optimal compromise between response mitigation and limited pendulum amplitudes.

Floating Offshore Energy Devices   Materials Research Forum LLC
Materials Research Proceedings **20** (2022) 39-46   https://doi.org/10.21741/9781644901731-6

Consistent modal decomposition

As shown in Fig. 1 the pendulum absorber is attached to the structure at two points (dofs j and k). In the present approach, the relative absorber motion is defined as the displacement relative to fully restraining the damper deformation, and thus rigidly linking the pendulum mass in Fig. 1 to the tower wall. The transformation that secures the correct relative absorber motion is

$$v = u_a - \mathbf{d}^T \mathbf{u}_s \tag{6}$$

Elimination of the absolute absorber displacement u_a by the relative absorber motion v in (6) gives the modified equations of motion

$$\left(\mathbf{M}_s + m_a \mathbf{d}\mathbf{d}^T\right)\ddot{\mathbf{u}}_s + \left(\mathbf{K}_s + k_a \Delta\mathbf{d}\Delta\mathbf{d}^T\right)\mathbf{u}_s + m_a \mathbf{d}\ddot{v} + k_a \Delta\mathbf{d}v = \mathbf{f}_s$$
$$m_a \ddot{v} + c_a \dot{v} + k_a v + m_a \mathbf{d}^T \ddot{\mathbf{u}}_s + k_a \Delta\mathbf{d}^T \mathbf{u}_s = 0 \tag{7}$$

where $\Delta\mathbf{dd} = \mathbf{d} - \mathbf{b}$ represents the difference in the connectivity vectors, that vanishes for the classic TMD ($\Delta\mathbf{d} = \mathbf{0}$).

This eigenvalue problem for this alternative representation includes the absorber mass as

$$\left(\mathbf{K}_s - \omega_s^2 (\mathbf{M}_s + m_a \mathbf{d}\mathbf{d}^T)\right)\bar{\mathbf{u}}_s = \mathbf{0} \tag{8}$$

with ω_s being the natural frequency for the structure with the augmented mass matrix. An improved representation could be obtained by also including the stiffness term in (7a), which however contains the – to be determined – absorber stiffness k_a. The structural displacement is now expressed by the expansion

$$\mathbf{u}_s = \sum_{j=1}^{N} \frac{\bar{\mathbf{u}}_j}{v_j} p_j \quad , \quad v_j = \mathbf{d}^T \bar{\mathbf{u}}_j \tag{9}$$

with p_j as the modal coordinate, N being the number of terms included and v_j representing the modal displacement at dof j where the dashpot in Fig. 1 is attached. Substitution of (9) into (7) followed by pre-multiplication with $\bar{\mathbf{u}}_s^T/v_s$ gives the following coupled modal equations for the targeted mode shape $j=s$,

$$m_s \ddot{p}_s + k_s p_s + k_a \frac{\Delta v_s^2}{v_s^2} p_s + m_a \ddot{v} + k_a \frac{\Delta v_s}{v_s} v = f_s$$
$$m_a \ddot{v} + c_a \dot{v} + k_a v + m_a \ddot{p}_s + k_a \frac{\Delta v_s}{v_s} p_s = 0 \tag{10}$$

In these modal equations, the change in modal connectivity is defined as $\Delta v_s = \Delta\mathbf{d}^T \bar{\mathbf{u}}_s$, while the interaction with other vibration modes is simply neglected in the supplemental (third) stiffness term in (10a) and the (two last) coupling terms in (10b).

The modal truncation creates the difference between the various modelling methods, since as little dynamics as possible should be neglected when truncating the series terms. In the present representation (6) relative to the damper deflection, the truncation is considered robust as it contains the limiting cases with vanishing and infinite absorber damping, whereby the absorber mass is included in the mode shapes by the eigenvalue problem (7). The representation in (6) thus

Floating Offshore Energy Devices
Materials Research Proceedings 20 (2022) 39-46

Materials Research Forum LLC
https://doi.org/10.21741/9781644901731-6

implies that the damping notoriously only appears in the absorber equation (10b), whereby it must not be omitted in the structural equation (10a).

Absorber calibration

The modal equations in (10) are expressed in the frequency domain with angular frequency ω, whereby variable p_s and v in the following denote the associated steady-state amplitudes. The modal equations can then be expressed as

$$
\begin{aligned}
\left(-\xi^2 + 1 + \kappa\Delta\gamma^2\right)p_s + \left(-\xi^2\mu + \kappa\Delta\gamma\right)v &= f_s/k_s \\
\left(-\xi^2\mu + i\xi\beta + \kappa\right)v + \left(-\xi^2\mu + \kappa\Delta\gamma\right)p_s &= 0
\end{aligned}
\tag{11}
$$

in which the non-dimensional frequency is defined as $\xi = \omega/\omega_s$, while the mass, stiffness and damper ratios

$$
\mu = \frac{m_a}{m_s} \quad , \quad \kappa = \frac{k_a}{k_s} \quad , \quad \beta = \frac{c_a}{\sqrt{m_s k_s}}
\tag{12}
$$

Are defined relative to the modal mass and stiffness

$$
m_s = \frac{\bar{\mathbf{u}}_s^T(\mathbf{M}_s + m_a\mathbf{d}\mathbf{d}^T)\bar{\mathbf{u}}_s}{v_s^2} \quad , \quad k_s = \frac{\bar{\mathbf{u}}_s^T\mathbf{K}_s\bar{\mathbf{u}}_s}{v_s^2}
\tag{13}
$$

The difference in modal deflection is in (11) represented by the relative increment $\Delta\gamma = \Delta v_s/v_s$. The characteristic equation follows from (11) as

$$
\xi^4 - \xi^2\frac{\mu + \kappa\left(1 + \mu((1 - \Delta\gamma)^2 - 1)\right)}{\mu(1 - \mu)} + \frac{\kappa}{\mu(1 - \mu)} + i\xi\frac{\beta}{\mu(1 - \mu)}\left(-\xi^2 + 1 + \kappa\Delta\gamma^2\right) = 0
\tag{14}
$$

The desired calibration with equal damping in the two modes associated with the targeted vibration form is in [6, 8, 9] obtained by comparison with the generic quartic polynomial

$$
\xi^4 - \xi^2\left(2 + 4\chi^2\right)\xi_0^2 + \xi_0^4 + i2\sqrt{2}\chi\xi_0\xi\left(-\xi^2 + \xi_0^2\right) = 0
\tag{15}
$$

Comparison of the last parentheses in (14) and (15) defines the reference frequency ratio as $\xi_0^2 = 1 + \kappa\Delta\gamma^2 \approx 1$, whereby comparison of the constant terms defines the stiffness ratio

$$
\kappa = \mu(1 - \mu)
\tag{16}
$$

which corresponds to the result in [9] for a classic TMD. The attainable damping is subsequently determined by the parameter

$$
4\chi^2 = \frac{\mu}{1 - \mu}\left(1 + (1 - \mu)\left((1 - \Delta\gamma)^2 - 1\right)\right)
\tag{17}
$$

For the classic TMD with $\Delta\gamma = 0$, the parenthesis in (17) becomes unity, whereby the correction for finite values of $\Delta\gamma$ appears because of the double attachment to dofs j and k in Fig. 1(c). The

Floating Offshore Energy Devices   Materials Research Forum LLC
Materials Research Proceedings 20 (2022) 39-46   https://doi.org/10.21741/9781644901731-6

explicit solution in (17) is simplified because of the coordinate shift in (6), whereby m_a is contained in the eigenvalue problem (7).

By comparison of the common coefficient to the odd-power terms in the characteristic equations (14) and (15), the damper ratio is finally obtained as

$$\beta = \sqrt{2\mu^3(1-\mu)\left(1 + (1-\mu)\left((1-\Delta\gamma)^2 - 1\right)\right)} \tag{18}$$

in which the correction coefficient from (17) is again recognized. This concludes the pendulum absorber tuning, which provides the blue curves in Fig. 2, comprising a suitable compromise between a flat response curve in (a) without any overshoot in (b). It should be noted that the classic TMD stiffness calibration in (5a) gives almost the same mitigation properties, when (5b) is replaced by the corrected damper ratio in (18).

The offshore wind turbine

This section provides the main information about the numerical offshore wind turbine model used to generate the results in Fig. 2. The geometry is shown in Fig. 3 and overall data is provided in Table 1. The top node mass includes the transverse and rotational inertia from the Rotor Nacelle Assembly ($M_{RNA} = 450 \cdot 10^3$ kg and $J_{RNA} = 120 \cdot 10^6$ kgm^2). The soil foundation is modelled by a Winkler spring layer with distributed stiffness $k_s = 200 \cdot 10^6$ (N/m)/m along the bottom $h_{soil} = 40$m, while water level is at $x = h_{sea} = 68$m (corresponding to a water depth of 28m).

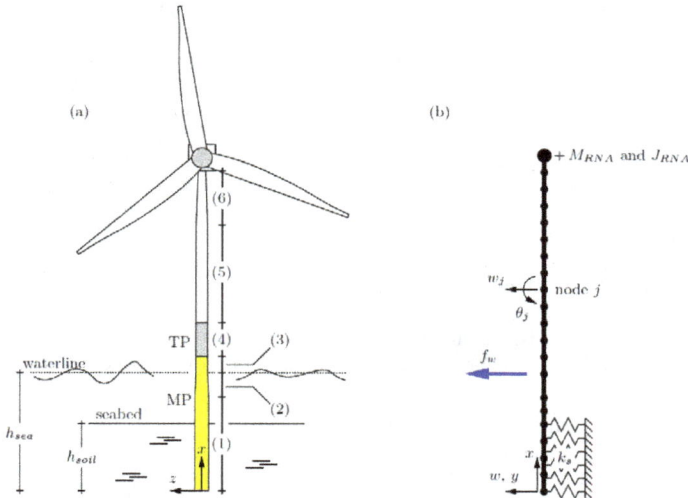

Figure 3: (a) Offshore wind turbine (owt) with data for section (1) to (6) provided in Table 1. The beam element model for the owt is shown in (b), with translation and rotational inerti from the Root Nacelle Assembly (RNA) added to the two top dofs. The model is supported by a Winkler foundation with stiffness k_s.

The owt (monopole + tower) model is discretized by 20 plane beam elements with two nodal dofs (transverse displacement and rotation). The material is steel and as geometric stiffness is neglected, the fundamental natural frequency $\omega_0 = 1.50$ rad/s without the pendulum absorber is slightly larger than in practice. A pendulum mass of $m_a = 10$ ton is assumed, which corresponds to a mass ratio of $\mu_0 = 1.54\%$ for the fundamental mode shape with respect to the top transverse displacement. As the length of the pendulum is approximately 22m, it is assumed in this example to simply span the top three beam elements, with attachment dofs in Fig. 1(c) given as $j = 39$ and $k = 45$. In reality, the span length should of course be re-adjusted according to the obtained absorber stiffness k_a. The harmonic load f_w – used to generate the frequency curves of Fig. 2 – is placed locally at sea level $x = h_{sea}$, as shown in Fig. 3(b).

Table 1: Section properties for offshore wind turbine model in Fig. 3.

Section	Height [m]	Outer diameter [mm]	Wall thickness [mm]
Monopile			
(1) Bottom section	55	8200	75
(2) Conical section	13	8200 → 6500	85
(3) Top section	10	6500	90
(4) Transition piece	19	6500	92
Tower			
(5) Bottom section	57	6500	55
(6) Conical section	35	6500 → 4200	30

The calibration based on the proposed modal representation gives a mass ratio $\mu = 0.79\%$, thus slightly smaller than $\mu_0 = 0.80\%$ for the TMD calibration based on a TMD mass attachment at the lower dof $j = 39$. The difference occurs because of the addition of the pendulum mass m_a to the mass matrix in (8). The pendulum stiffness $k_a = 22.1$ kN/m is slightly smaller than the value 22.1 kN/m obtained for a TMD placed locally at the top dof $k = 45$. This change in stiffness creates the visible inclination of the red curve in Fig. 2(a) and its corresponding overshoot in Fig. 2(b). The actual absorber damping coefficient is found to be $c_a = 2.59$ kNs/m, which is somewhat larger than the simplified TMD solution of 1.87 kNs/m, associated with the placement at dof $j = 39$. The present calibration method therefore consistently incorporates that the pendulum absorber acts on the tower at two different positions, whereby it basically determines an absorber stiffness associated with the top deflection and a damper value that is proportional to the deflection at damper position. It should be emphasized that the classic TMD tuning procedure is incapable of accounting for this effect as it notoriously assumes a single point of attachment of the classic TMD.

Summary
A consistent absorber calibration procedure is devised for an absorber with two different points of attachment. The calibration procedure is illustrated for a simple offshore wind turbine model with a pendulum-type absorber attached to the top tower position and with the dashpot acting horizontally between tower wall and pendulum mass. The proposed calibration formulae are seen to provide a good compromise between effective response mitigation and limited pendulum vibration amplitudes. The concept can be generalized to flexible absorbers with no common or even with distributed points of attachment to the host structure.

References

[1] S. Elias, V. Matsagar, Research developments in vibration control of structures using passive tuned mass dampers, Annual Reviews in Control 44 (2017) 129-156. https://doi.org/10.1016/j.arcontrol.2017.09.015

[2] S. Park, M.A. Lackner, P. Pourazarm, A.R. Tsouroukdissian, J. Cross-Whiter, An investigation on the impacts of passive and semiactive structural control on a fixed bottom and a floating offshore wind turbine, Wind Energy (2019). https://doi.org/10.1002/we.2381

[3] L.K. Wang, W.X. Shi, X.W. Li, Q.W. Zhang, Y. Zhou, An adaptive-passive retuning device for a pendulum tuned mass damper considering mass uncertainty and optimum frequency, Structural Control and Health Monitoring 26 (2019) e2377. https://doi.org/10.1002/stc.2377

[4] S. Colwell, B. Basu, Tuned liquid column dampers in offshore wind turbines for structural control, Engineering Structures 31 (2009) 358-368. https://doi.org/10.1016/j.engstruct.2008.09.001

[5] J.P. Den Hartog, Mechanical Vibrations, fourth ed., Dover, New York, 1985.

[6] S. Krenk, Frequency analysis of the tuned mass damper, Journal of Applied Mechanics 72 (2005) 936-942. https://doi.org/10.1115/1.2062867

[7] M. Zilletti, S.J. Elliott, E. Rustighi, Optimisation of dynamic vibration absorbers to minimise kinetic energy and maximise internal power dissipation, Journal of Sound and Vibration 331 (2012) 4093-4100. https://doi.org/10.1016/j.jsv.2012.04.023

[8] S. Krenk, J. Høgsberg, Tuned resonant mass or inerter-based absorbers: unified calibration with quasi-dynamic flexibility and inertia correction, Proceedings of the Royal Society A - Mathematical, Physical and Engineering Sciences 472 (2016) paper no. 20150718 (23pp). https://doi.org/10.1098/rspa.2015.0718

[9] D. Hoffmeyer, J. Høgsberg, Calibration and balancing of multiple tuned mass absorbers for damping of coupled bending-torsion beam vibrations, submitted for publication (2019). https://doi.org/10.1115/1.4046752

Floating Offshore Energy Devices
Materials Research Proceedings 20 (2022) 47-57

Materials Research Forum LLC
https://doi.org/10.21741/9781644901731-7

Machine Learning for Wind Turbine Fault Prediction through the Combination of Datasets from Same Type Turbines

Cristian Bosch[1,a*] and Ricardo Simon Carbajo[1,b]

[1] Ireland's Centre for Applied AI (CeADAR), School of Computer Science, University College Dublin, Ireland

[a]cristian.boschserrano@ucd.ie, [b]ricardo.simoncarbajo@ucd.ie

Keywords: Predictive Maintenance, Wind Turbine, Machine Learning, Artificial Intelligence, Optimal Transport

Abstract. Early fault detection in wind turbines is key to reduce both costs and uncertainty in the generation of energy and operation of these structures. The isolation of many wind farms, especially those offshore, makes scheduled maintenance very costly and on many occasions inefficient. In addition, the downtime of these structures is typically long and a predictive solution is much needed to 1) help prepare for the maintenance procedure beforehand, for instance to avoid delays when waiting for the required resources and components for maintenance to be available and, 2) avoid the possibility of more destructive system failures. Predicting failures in such complex systems requires modeling of multiple components in isolation and as a whole. Physics-based and data-based models are used for this purpose, which have been proven useful in this regard. Specifically, Machine Learning algorithms are proven to be a valuable resource in a wide range of problems in this industry, however a solution capable of accurately predicting the range of faults of a particular type of wind turbine is still a challenge. In this paper, we will introduce the capabilities of machine learning for wind turbine fault prediction, as well as a technique to predict different types of faults. We will compare the performance of two well established machine learning algorithms (namely K-Nearest Neighbour and Random Forest classifiers) on real wind turbine data which have produced great levels of prediction accuracy. We also propose data augmentation methods to help enhance the training of ML models when wind turbine data is scarce by merging data from turbines of the same type.

Introduction

According to WindEurope [1], wind energy accounts for the second largest source of power for the European Union (EU), with a 18.8% of its capacity, after natural gas. Ireland accounts for the 3.5% of the EU's cumulative capacity, that covers a 28% of the energy demand of the country. In this context, it is paramount to remark that the maintenance cost of a wind turbine can range from a fifth [2] to a third [3] part of the levelized cost of energy. Being wind energy a mature but rising technology, solving this issue is a top priority in order to contribute to its rapid, sustainable and cost-efficient adoption. Thus, the area of predictive maintenance for these structures has become fundamental and different approaches aim for the goal of reducing downtime, lessen the damage and prolonging wind turbine lifetime.

While physics-based modeling systems exists, our purpose is to approach the problem through the application of several Machine Learning (ML) algorithms on the data collected by the Condition Monitoring System (CMS) through the SCADA system from several wind turbines. For this work, we have obtained data from a set of onshore Siemens SWT-2.3-101 wind turbines. The goal is, using historical data of previous registered faults, predicting the failure of different parts

Floating Offshore Energy Devices
Materials Research Proceedings **20** (2022) 47-57

Materials Research Forum LLC
https://doi.org/10.21741/9781644901731-7

of the turbine giving a general forecast of downtime with an anticipation of at least 24 hours as to have maintenance scheduled in a reliable manner.

The paper will explain the steps involved in data preparation, faults and features analysis, the rationale to establish the minimum time to predict a fault and the overall modeling setting to train and test the system. It will also include a statistical analysis to validate the combination of datasets from turbines of the same type as a data augmentation technique. After training and testing models using a variety of machine learning algorithms, including neural networks, the two most promising machine learning methods will be presented and compared. The first algorithm is the K-Nearest Neighbour classifier (KNN) [4], while the second algorithm is the Random Forest classifier [5], which is an ensemble of decision trees. A range of metrics (precision, recall, f1-score and support) will be presented to quantify and explain the accuracy and capabilities of the proposed predictive models applied to wind turbines and their impact.

The paper is structured as follows: next section presents the state of the art in the area of wind turbine fault prediction using machine learning. Subsequently, the methodology of our approach is explained, where the analysis of the wind turbine data is presented, including faults and features, and the machine learning classifiers introduced together with the methodology of their application to the data model created. Results are then provided and discussed in terms of different prediction accuracy metrics for both the modelling using a dataset from one turbine and augmenting the dataset by combining datasets from two turbines of the same type. Finally, our conclusions and future research avenues are highlighted.

State of the Art

Wind turbines are composed of different rotating parts that undergo an intensive performance through its lifetime. Condition Monitoring Systems (CMSs) are common in the current industry and use a collection of sensors that monitor the state of the different parts of the turbine in real time. A wide range of sensors are used to measure: vibrations, oil levels and temperature, thermographic analysis, crack detection, strain, acoustic analysis, electrical conditions, signal and performance monitoring [6]. These measurements are combined by CMSs for the monitoring of specific subsystems of the wind turbine.

There are approaches in the literature which focus on modelling the operation of wind turbine components using physics models while enhancing these models with data from the wind turbine collected through CMSs to create a hybrid approach. However, the challenge in this paper is to only exploit data (both from the operations of the wind turbine and auxiliary data) to model the behavior of similar aspects of wind turbines so models can be transfered to cover a larger range of wind turbine types.

Focusing on exploiting these data using ML algorithms, the fault detection problem can be tackled with two different strategies: i) studying the normal regime of performance and detecting anomalies and ii) analyzing the time periods prior to a fault to build models that can anticipate faulty behavior.

With respect to anomaly detection, modelling the normal performance of a wind turbine has the advantage of using most of the CMS data collected, since turbine datasets are greatly imbalanced towards the normal regime. We could qualify these approaches as semi-supervised algorithms, considering that faults and their adjacent data in time are purposely removed before training the models. An elegant solution that follows this strategy is using autoencoders. An autoencoder is a deep Neural Network (NN) built symmetrically to filter the relevant features of data and learn the relationships of the different variables under certain conditions. In other words, the NN learns how the wind turbine works in essence, by filtering out noise. Autoencoders have obtained good results in early detection of faults and allowed for the discrimination of the mechanical parts implied in the fault state [7]. The literature contains different approaches for anomaly detection too. For example, other types of NN architectures can be trained with data from the normal regime of a wind turbine and learn to predict the expected power output at any given moment which, compared against the real output, is able to pinpoint defective behavior. This behavior is then traced to the parts through a Principal Component Analysis (PCA), which is a used to reduce the dimensionality of a dataset [8]. Classification methods are also used where the normal behavior periods are constrained to be far away from a fault and have been proven to be an adequate mechanism to select which SCADA features should be considered for fault detection [9].

In relation to the analysis of historical faults, we can find a wide range of techniques too. The different classifiers described in the literature are examples of supervised training, which means that every data entry is labeled. Other approaches which do not use analytics of the SCADA data, employ visual inspection of the turbine through a drone and apply Convolutional Neural Networks (CNNs) to process the images and detect common external damages such as erosion or missing teeth in the vortex generator [10]. The image datasets, like the SCADA data, have in common that they are highly imbalanced, which requires specific architectures for the CNNs [11]. Moving back to turbine sensor data, multiclass classification has been attempted on simulations of turbine data through Support Vector Machines (SVMs), succeeding in the isolation of different faults [12]. SVMs have been very popular for fault detection in previous years, however decision trees plus boosting techniques have gained relevance recently. Random Forest and XGBoost classifiers have been used to study the relevance of features and detect faults in other wind turbine models [13]. Signal analysis on the currents of the Double-Fed Inductor Generator (DFIG) have also been performed. The current of the DFIG would experience interference due to the vibrations of a faulty gearbox, and thus, autoencoders and NN classifiers are able to predict impending faults when applied to this signal [14].

This paper focusses on modelling real SCADA data from wind turbines of the same type via the application of a large number of ML classifiers and the use of a novel data augmentation technique.

Methodology

The work presented in this paper explored a large range of ML algorithms on real data from a wind turbine and identified two algorithms providing the best predictive performance (K-NN and Random Forest). However, one of the main problems when dealing with this type of data is that training one model for each wind turbine is not generalizable and is limited by the size of the dataset and the scarcity of faulty data. To tackle such problem, we propose a novel system to combine datasets coming from turbines of the same type. We analyse the similarities of the datasets through PCA and calculate the statistical distance between both datasets and then use Optimal Transport (OT) [15] to transform the probability distribution of one dataset into another, such that we can augment the data by combine both datasets.

<u>Data description</u>. We have obtained data from two Siemens SWT-2.3-101 turbines (namely Turbine 1 and 2) over the same period and belonging to the same wind farm. These data are collected every ten minutes and comprise almost five years of SCADA aggregated recording. We have selected the features presented in Table 1 for training our models, as they present the most useful information recorded. As a remark, only nine of these features are free from strong correlations. However, we consider interesting to include them all as these correlations may experience certain deviations in the proximity of faults.

Floating Offshore Energy Devices

Materials Research Proceedings **20** (2022) 47-57

Materials Research Forum LLC

https://doi.org/10.21741/9781644901731-7

Table 1. Features from our dataset used for modelling.

Feature	Type	Feature	Type
Wind Speed [m/s]	Mechanical	Blade Angle (Pitch Pos.) C [º]	Mechanical
Nacelle Position [º]	Mechanical	Blade Pressure (Hydraulic) [bar]	Mechanical
Rear Bearing Temp. [º]	Mechanical	Power [kW]	Power and Electricity
Ambient Temp. [º]	Mechanical	Voltage L1 [V]	Power and Electricity
High Speed Bearing Temp. [º]	Mechanical	Voltage L2 [V]	Power and Electricity
Gear Bearing 1 Temp. [º]	Mechanical	Voltage L3 [V]	Power and Electricity
Gear Bearing 2 Temp. [º]	Mechanical	Current L1 [A]	Power and Electricity
Rotor Speed [RPM]	Mechanical	Current L2 [A]	Power and Electricity
Generator [RPM]	Mechanical	Current L3 [A]	Power and Electricity
Blade Angle (Pitch Pos.) [º]	Mechanical	WTOperation State	Status Flag
Blade Angle (Pitch Pos.) A [º]	Mechanical	Errorcode	Status Flag
Blade Angle (Pitch Pos.) B [º]	Mechanical	WpsStatus	Status Flag

The data are cleaned of null values and labelled before training ML models. Since we want to detect fault events with anticipation to schedule an emergency maintenance, the criteria used for labelling is to consider as "prefault" any data prior to a recorded fault that causes downtime. As a commitment between the balance of data labels and the necessity of the industry to schedule the maintenance in advance, we have decided to consider "prefault" data all the entries belonging to a period 36 hours before the fault event. We will find downtime periods that are induced due to human intervention or automatic stops, such as scheduled maintenance. We use data from four days (empirical threshold) before any of these faults happen to fit a scaler in the case of K-Nearest Neighbors model training. This choice of threshold is taken to isolate the data representing the normal regime of operation of the turbine as much as possible, forcing the scaler to work within the normal operation range for any of its features. Using points that include faulty behavior, as they tend to be extremal (a fault event can generate outliers that we do not want to remove but we need to avoid transforming data with them), could make us lose granularity in the data due to normalizing with a value that is too high. Principal Component Analysis has been applied to these normalized datasets with the purpose of visualizing and understanding the impact of each feature in its behavior. An interesting conclusion obtained by the combination of normalization and PCA is that both turbines work in similar intervals for every feature and thus, knowing that their behavior is equivalent, we can devise a data augmentation strategy to combine their records.

A large range of different fault types has been found in our datasets. We have 953 penalizing downtime events for the first turbine which, as expected, do not exhibit similar behavior and, in

Floating Offshore Energy Devices
Materials Research Proceedings **20** (2022) 47-57

Materials Research Forum LLC
https://doi.org/10.21741/9781644901731-7

many cases, it is much less than 24 hours. Approximately 10% of these events are human induced stops. However, since they are not only originated by a scheduled maintenance but also from reactive condition monitoring decisions, we will include them in our model assuming the associated error. Similarly, the second turbine presents 981 downtime events where approximately 8.5% of them are due to human intervention.

This is one of the reasons why we are choosing to take a supervised classification approach and not a time series analysis based on error code segregation, since the latter will have to be performed with the disadvantage of errors clearly misrepresented in the dataset compared to those that appear frequently and, thus, without enough data to train a model properly.

Data Augmentation via Optimal Transport. As we stated before, the two wind turbines have a nearly identical representation in the feature space after performing PCA on them. We present the analysis on the data extraction of what we have defined as normal behavior. Figure 1 shows the PCA reduction to two dimensions of both turbines, presenting a striking similarity.

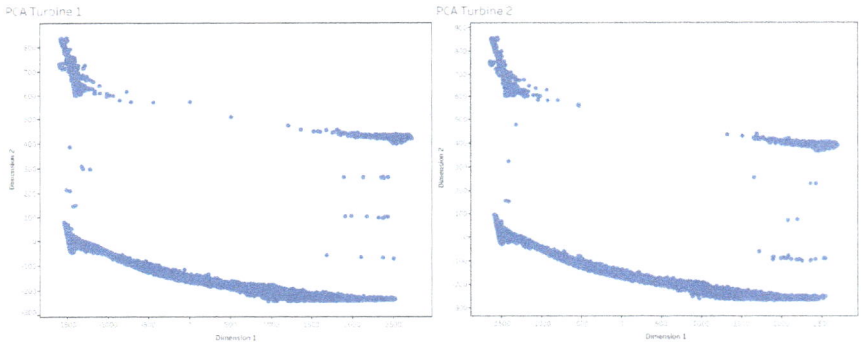

Figure 1. Two principal dimensions of the datasets belonging to the two turbines after applying PCA to the data characterizing normal behavior of the wind turbines.

Nevertheless, the combination of the datasets must be optimized to evaluate their statistical distance and to transfer one distribution closer to the other. We have computed the Earth's Mover Distance (EMD), equivalent to the Wasserstein's Distance for the two distributions, on each of the features we are going to train, to minimize it for these two datasets. The Wasserstein distance between distributions is defined by:

$$W_p(a,b) = \left(\min_{\gamma \in \mathbb{R}_+^{m \times n}} \Sigma_{i,j} \gamma_{i,j} \|x_i - y_j\|_p \right)^{\frac{1}{p}} \text{ such that } \gamma 1 = a; \ \gamma^T 1 = b; \ \gamma \geq 0 \tag{1}$$

We have used the Python Optimal Transport library [16] for this purpose. The original value for the squared Euclidean average of every feature is EMD = 0.001639. This has been computed by scaling every turbine dataset independently. We take every feature column of the dataset and cut it in bins (with a maximum number of bins of 20), comparing feature by feature the data of the

Floating Offshore Energy Devices
Materials Research Proceedings **20** (2022) 47-57

Materials Research Forum LLC
https://doi.org/10.21741/9781644901731-7

turbines. The centers of the bins are used to compute the distance matrix with squared Euclidean distance metric, whereas we compute the EMD with the relative frequency of samples in each bin.

The approach for achieving optimal transport of the data consists of declaring the mean and the variance of each feature distribution in the scaler for the second turbine as hyperparameters and then optimize their value in order to minimize this distance. The StandardScaler class [17] is used as defined in Eq. 2. The final value after this process, where the second dataset is scaled before merging their data, gives a EMD = 0.001567. This value is close to the previously unoptimized EMD value and shows that both datasets are statistically close and can be transported and combined to augment the dataset for ML modelling.

$$z = \frac{x-\mu}{\sigma} \quad \text{such that } x = \text{sample; } \mu = \text{mean; } \sigma = \text{standard deviation} \tag{2}$$

After this optimized data transfer, we split our new dataset respecting the time series nature of each dataset and then the training, validation and tests sets that will be computed separately are merged for building the final model.

Classification Models. The data is labeled according to the three following classes: Normal, Prefault and Fault. The idea is that, after training our model, new SCADA observations will be classified according to one of them and will be used to make accurate predictions of an undefined fault. A set of ML algorithms has been tested, however only two are presented which exhibit the best accuracy.

The first ML algorithm is the K-Nearest Neighbors (KNN) classifier. This algorithm has a straightforward training method. It generates a space with a dimensionality equal to the number of features used for training. Training consists of mapping every data point into this space. Prediction is performed by locating the new point in this space and then retrieving the label of its K closest points according to a determined geometrical distance and a weight assigned through a function that normally will depend on this distance. This algorithm and its parameters (number of neighbors, distance metric, leaf size, power parameter for the metric and weights) are easy to understand but it was traditionally qualified as resource-consuming, which is no longer a problem with current computers. For the correct performance of this estimator, we will scale the data using Scikit-Learn's StandardScaler, fitting the normal regime of the turbines as previously explained. We will compare the performance on each individual turbine and after the datasets belonging to different turbines are merged following the optimal transport strategy previously outlined, with the intent of building a global model for this type of turbine.

The second ML algorithm is the Random Forest classifier. The Random Forest algorithm is an ensemble of decision trees that selects the best candidate after a majority voting process. A decision tree is an algorithm that splits data into subsets according to decision nodes that are based in the values of different features. A decision tree will grow, node by node, branching as its being trained with data to reproduce the expected output, a category or a continuous variable depending on the nature of the problem, i.e., classification or regression. The number of hyperparameters that define the algorithm's behavior is larger than for the KNN model, since its depth (the quantity of features that form the pool where one is chosen for a tree node) and the way the decision is taken (symmetrically or not with respect to the interval of the data), amongst others, can be determined. The Random Forest will be built by a determined number of decision trees and, as an ensemble, it will perform better than a single estimator with a considerable improvement when the dataset becomes larger [18]. Once the model is trained, predictions will be based on a majority vote system

Floating Offshore Energy Devices

Materials Research Forum LLC

Materials Research Proceedings 20 (2022) 47-57

https://doi.org/10.21741/9781644901731-7

for the individual estimators. This estimator does not require any normalization as it makes the decisions based on a set value or category. However, we will provide the data scaled in the case where we merge the datasets.

In order to find the best training hyperparameters for these two models, we will perform a random search using RandomSearchCV from scikit-learn and use the KNN and Random Forest functions in this library, with the use of a validation set. Training, validation and test sets are fairly balanced with regards to their labels for both turbines (the merged dataset will be balanced as a consequence). For the random search hyperparameter values are sampled from different distributions according to them being continuous, discrete numeric or categorical variables to find those that fit best the dataset. The metrics defining the performance of our model will be computed on the validation set. We will present our results by the application of the trained model to the test set, which gives the final metrics defining the performance of the model.

Results

As our purpose is to evaluate our proposed data augmentation strategy, we will present the results of applying these classifiers first to the dataset of each turbine and then to the augmented dataset that combines data from both turbines.

The hyperparameter optimization on both classifiers is performed using "f1-score" as the metric. This choice is motivated by the intent of making the model focus on maximizing the harmonic average of "precision" and "recall". "Precision" is defined as the ratio of the correct predictions over all the predictions of a label and "Recall" is the ratio of correct predictions over all occurrences of a label.

Experiments with Turbine 1 Data. The random search applied to find the best hyperparameters for fitting the KNN classifier produces the following setting: i) four neighbors, ii) $p = 1$, which means the Manhattan distance is being used (this is, as if the space was gridded and the distances were computed by counting the sides of the squares), iii) weighted according to the distance and iv) leaf size $= 8$, which is useful for tree building depending on the algorithm (which we have set as automatic). Using these parameters, our pre-fault detection, considering as pre-fault all the data belonging to the last 36h before a fault causes downtime, produces the accuracy results given in Table 2. In the table we can see that using this model the results for the test set in terms of recall and precision indicate that: we can predict 41% of the faults (i.e., Prefaults) and that 94% of the time when our predictor indicates there is a fault within the next 24 hours, the predictor will be correct.

Table 2. KNN metrics for the Turbine 1 in the test set.

Label / Average	Precision	Recall	F1-score	Support
Normal-0	0.64	0.97	0.77	29687
Prefault-1	0.94	0.41	0.57	27625
Fault-2	0.88	0.73	0.80	239
Accuracy (micro)			0.70	57551
Macro	0.82	0.71	0.72	57551
Weighted	0.78	0.70	0.68	57551

Floating Offshore Energy Devices
Materials Research Proceedings **20** (2022) 47-57

Materials Research Forum LLC
https://doi.org/10.21741/9781644901731-7

Moving now to the Random Forest classifier, the best model we find after the random search is characterized by the following hyperparameters: i) bootstrap = True, this means that a part of the dataset is used for building each tree, ii) max_depth = 40, this defines how long the tree can extend, iii) max_features = auto, the number of features checked before doing any split, iv) min_samples_leaf = 3, minimum number of samples required to be in a leaf node, v) min_samples_split = 9, which defines the number of samples required to split internal nodes, and vi) number of estimators = 409, which represents the number of trees forming the ensemble. The metrics on the test set of the Turbine 1 are shown in Table 3. As we can see, our metric of interest has a slightly superior precision (96%) and a slightly inferior recall (43%), showing no considerable improvement from the KNN result.

Table 3. Random Forest classifier metrics for the Turbine 1 in the test set.

Label / Average	Precision	Recall	F1-score	Support
Normal-0	0.60	0.99	0.75	29687
Prefault-1	0.96	0.43	0.48	27625
Fault-2	0.97	0.72	0.82	239
Accuracy (micro)			0.66	57551
Macro	0.84	0.67	0.68	57551
Weighted	0.78	0.66	0.62	57551

Experiments with Turbine 2 Data. As we did with the Turbine 1 dataset, we will begin by presenting the best KNN model found by the random search hyperparameter optimization. Its hyperparameters are: i) three neighbors, ii) p = 1 (Manhattan distance), iii) weighted according to the distance and iv) leaf size = 27. The performance results of this model in the test set are presented in Table 4. The results are comparable to those of Turbine 1. The label of interest, which is Prefault, has a detection precision of 92% and the model can recall a 44% of the Prefaults.

Table 4. KNN metrics for the Turbine 2 in the test set.

Label / Average	Precision	Recall	F1-score	Support
Normal-0	0.68	0.97	0.80	29775
Prefault-1	0.92	0.44	0.60	24999
Fault-2	0.84	0.78	0.81	246
Accuracy (micro)			0.73	55020
Macro	0.81	0.73	0.74	55020
Weighted	0.79	0.73	0.71	55020

Regarding the Random Forest classifier, the best model yields the metrics presented in Table 5. The hyperparameters that optimize the f1-score are: i) bootstrap = False, now we are using the whole dataset for building trees, ii) max_depth = 26, iii) max_features = log2, this is a common way of deciding the number of features to include before splitting, iv) min_samples_leaf = 4, v) min_samples_split = 10 and vi) number of estimators = 303. For this turbine, Random Forest penalizes the recall over the precision, which makes the KNN model a more balanced choice for forecasting faults.

Table 5. Random Forest classifier metrics for the Turbine 2 in the test set.

Label / Average	Precision	Recall	F1-score	Support
Normal-0	0.63	0.99	0.77	29775
Prefault-1	0.96	0.35	0.52	24999
Fault-2	0.93	0.67	0.78	246
Accuracy (micro)			0.69	55020
Macro	0.84	0.67	0.69	55020
Weighted	0.79	0.69	0.65	55020

Experiments with Combination of Turbine 1 and 2 Data. Following the strategy outlined in the "Data Augmentation via Optimal Transport" subsection, we augment our dataset by combining data from both wind turbines. Once we have this larger dataset, we run another random search for each of the classifier models.

Table 6 shows the best KNN model after the hyperparameters are optimized. These hyperparameter values are: i) four neighbors, ii) $p = 1$ (Manhattan distance), iii) weighted according to the distance and iv) leaf size = 14.

Table 6. KNN metrics for the combined dataset in the test set.

Label / Average	Precision	Recall	F1-score	Support
Normal-0	0.67	0.97	0.79	59462
Prefault-1	0.93	0.45	0.61	52624
Fault-2	0.83	0.72	0.77	485
Accuracy (micro)			0.73	112571
Macro	0.81	0.71	0.73	112571
Weighted	0.79	0.73	0.71	112571

As we stated before, KNN is a simple classification algorithm. Thus, even though we do not expect that the results massively improve, we can observe that the recall on the Prefault labelled data has increased as compared to modelling the turbines separately, thus improving the f1-score. Therefore, our data augmentation strategy yields favorable results.

Moving now to the Random Forest classifier, the optimal hyperparameters that our random search exploration of the hyperparameter space produced are: i) bootstrap = True., ii) max_depth = 60, iii) max_features = auto, iv) min_samples_leaf = 6, v) min_samples_split = 13 and vi) number of estimators = 96. The metrics are presented in Table 7. These results show the same trend that the Turbine 2 had when trained independently. The model tends to improve the precision over a more balanced approach with recall.

These results show the potential benefit of combining datasets for the same type of turbine through the proposed approach, especially when we combined more than two datasets; this is future work. Note that the application of the ML classifiers has not considered the time series nature of the data and the proximity of observations (every 10 minutes). Different dataset splitting strategies are being investigated along with regression algorithms using a combination of deep learning techniques which will be presented in future works.

Floating Offshore Energy Devices
Materials Research Proceedings **20** (2022) 47-57

Materials Research Forum LLC
https://doi.org/10.21741/9781644901731-7

Table 7. Random Forest metrics for the combined dataset in the test set.

Label / Average	Precision	Recall	F1-score	Support
Normal-0	0.62	0.99	0.76	59462
Prefault-1	0.96	0.34	0.51	52624
Fault-2	0.93	0.76	0.84	485
Accuracy (micro)			0.68	112571
Macro	0.84	0.70	0.70	112571
Weighted	0.78	0.68	0.64	112571

Conclusions

The main purpose of our publication has been accomplished. Data from wind turbines belonging to the same brand and model can potentially be combined with our proposed data augmentation strategy for more general model building.

As we have used very simple Machine Learning models, the results will not excel in forecasting capabilities, but define a route for future work with a deeper exploration of the time series properties of the data and advanced data augmentation through transfer learning techniques such as Generative Adversarial Networks, which are closely related to the optimal transport approach proposed.

Nevertheless, we have proven that after solving the optimal transport problem, the models do not lose predictability, which would be expected if the datasets had completely different projections in the feature space, but this predictability can be reinforced as the KNN model results show.

Regarding future work, a positive path towards the creation of solid extensively trained models has been laid out that could represent the behavior of an ideal turbine with its common errors at the model (or associated to specific turbine models), even mixing different wind farms and periods if the optimal transport dataset transformation is handled properly. This would be useful as customers could use their aggregated data (sometimes insufficient) to build their own maintenance management algorithms.

Acknowledgements

The authors thank Enterprise Ireland and the European Union's Horizon 2020 research and innovation programme for funding under the Marie Skłodowska-Curie grant agreement No. 713654.

References

[1] WindEurope; Wind energy in Europe in 2018 - Trends and statistic, 2019.

[2] Sara Verbruggen; Onshore Wind Power Operations and Maintenance to 2018. ENDSIntelligence, 2016.

[3] Feng, Y.; Tavner, P.; Long, H. Early experiences with UK round 1 offshore wind farms. Proc. Inst. Civ. Eng., 2010, 163, 167–181. https://doi.org/10.1680/ener.2010.163.4.167

[4] Cunningham, Padraig & Delany, Sarah; k-Nearest neighbour classifiers, 2007, Mult Classif Syst.

Floating Offshore Energy Devices
Materials Research Proceedings 20 (2022) 47-57

Materials Research Forum LLC
https://doi.org/10.21741/9781644901731-7

[5] J. J. Rodriguez, L. I. Kuncheva and C. J. Alonso; "Rotation Forest: A New Classifier Ensemble Method," in IEEE Transactions on Pattern Analysis and Machine Intelligence, 2006, vol. 28, no. 10, pp. 1619-1630. https://doi.org/10.1109/TPAMI.2006.211

[6] Z. Hameed, Y.S. Hong, Y.M. Cho, S.H. Ahn, C.K. Song; Condition monitoring and fault detection of wind turbines and related algorithms: A review, Renewable and Sustainable Energy Reviews, Volume 13, Issue 1, 2009, Pages 1-39B. https://doi.org/10.1016/j.rser.2007.05.008

[7] Hongshan Zhao, Huihai Liu, Wenjing Hu, Xihui Yan; Anomaly detection and fault analysis of wind turbine components based on deep learning network, Renewable Energy, Volume 127, 2018, Pages 825-834. https://doi.org/10.1016/j.renene.2018.05.024

[8] Mazidi, Peyman & Bertling Tjernberg, Lina & Sanz-Bobi, Miguel; Performance analysis and anomaly detection in wind turbines based on neural networks and principal component analysis, Conference: 12th Workshop on Industrial Systems and Energy Technologies, 2017.

[9] Felgueira, T., Rodrigues, S., Perone, C.~S., et al.; 2019, arXiv e-prints, arXiv:1906.12329.

[10] Shihavuddin, A & Chen, Xiao & Fedorov, Vladimir & Riis, Nicolai & Nymark Christensen, Anders & Branner, Kim & Bjorholm Dahl, Anders & Reinhold Paulsen, Rasmus; Wind Turbine Maintenance Cost Reduction by Deep Learning Aided Drone Inspection Analysis, 2019. https://doi.org/10.20944/preprints201901.0281.v1

[11] Anantrasirichai, Nantheera & Bull, David; DefectNET: multi-class fault detection on highly-imbalanced datasets, 2019, arXiv e-prints, arXiv:1904.00863. https://doi.org/10.1109/ICIP.2019.8803305

[12] Abderrahmane, Mokhtari \& Belkheiri, Mohammed; Fault Diagnosis of a Wind Turbine Benchmark via Statistical and Support Vector Machine. International Journal of Engineering Research in Africa, 2018, 37. 29-42. https://doi.org/10.4028/www.scientific.net/JERA.37.29

[13] D. Zhang, L. Qian, B. Mao, C. Huang, B. Huang and Y. Si; "A Data-Driven Design for Fault Detection of Wind Turbines Using Random Forests and XGboost,", 2018, IEEE Access, vol. 6, pp. 21020-21031. https://doi.org/10.1109/ACCESS.2018.2818678

[14] F. Cheng, J. Wang, L. Qu and W. Qiao; "Rotor current-based fault diagnosis for DFIG wind turbine drivetrain gearboxes using frequency analysis and a deep classifier," 2017, 2017 IEEE Industry Applications Society Annual Meeting, Cincinnati, OH, pp. 1-9. https://doi.org/10.1109/IAS.2017.8101844

[15] Villani, Cédric. Optimal transport: old and new. Vol. 338. Springer Science & Business Media, 2008.

[16] Rémi Flamary and Nicolas Courty, POT Python Optimal Transport library, Website: https://pythonot.github.io/, 2017.

[17] Pedregosa, F. and Varoquaux, G. and Gramfort, A. and Michel, V. and Thirion, B. and Grisel, O. and Blondel, M. and Prettenhofer, P.and Weiss, R. and Dubourg, V. and Vanderplas, J. and Passos, A. and Cournapeau, D. and Brucher, M. and Perrot, M. and Duchesnay, E.; Scikit-learn: Machine Learning in Python, Journal of Machine Learning Research, 2012, Volume 12, pages 2825-2830.

[18] Ali, Jehad, et al. Random forests and decision trees. International Journal of Computer Science Issues (IJCSI), 2012, vol. 9, no 5, p. 272.

Floating Offshore Energy Devices
Materials Research Proceedings 20 (2022) 58-65

Materials Research Forum LLC
https://doi.org/10.21741/9781644901731-8

Ensemble Empirical Mode Decomposition Based Deep Learning Model for Short-Term Wind Power Forecasting

Juan Manuel González-Sopeña[1,a*], Vikram Pakrashi[2,b] and Bidisha Ghosh[1,c]

[1]Department of Civil, Structural and Environmental Engineering, Trinity College Dublin, Ireland

[2]Dynamical Systems and Risk Laboratory, School of Mechanical & Materials Engineering, University College Dublin, Ireland

[a]gonzlezj@tcd.ie, [b]vikram.pakrashi@ucd.ie, [c]bghosh@tcd.ie

Keywords: Short-Term Wind Power Forecasting, Ensemble Empirical Mode Decomposition, Deep Learning, Prediction Intervals, Quantile Regression, Wind Power

Abstract. In the last few years, wind power forecasting has established itself as an essential tool in the energy industry due to the increase of wind power penetration in the electric grid. This paper presents a wind power forecasting method based on ensemble empirical mode decomposition (EEMD) and deep learning. EEMD is employed to decompose wind power time series data into several intrinsic mode functions and a residual component. Afterwards, every intrinsic mode function is trained by means of a CNN-LSTM architecture. Finally, wind power forecast is obtained by adding the prediction of every component. Compared to the benchmark model, the proposed approach provides more accurate predictions for several time horizons. Furthermore, prediction intervals are modelled using quantile regression.

Introduction

Wind energy provides an alternative source of electricity generation. Compared to traditional sources of energy, wind is a highly intermittent and volatile resource. Therefore, accurate wind power forecasts are fundamental for the proper operation of the grid [1], as well as maximizing results in the electricity market [2].

Many forecasting techniques have been proposed to forecast wind power generation [3]. They can be broadly divided into physical and statistical methods. The first approach makes use of meteorological information and the specific site conditions of the wind farm. Statistical models are usually built using historical data. Conventional statistical methods use time series modelling to predict future values of wind power output. For instance, [4] proposed a method based on wavelets and the improved time series method (ITSM) to forecast wind speed and wind power. An alternative statistical approach is to employ deep learning techniques as artificial neural networks (ANNs) [5, 6].

In addition, several models have been introduced to build prediction intervals (PIs). Quantile regression (QR) is a well-known technique characterized by its distribution-free approach [7]. Other non-parametric forecasting models are based on kernel density estimation [8, 9]. [10] employs the lower upper bound estimation (LUBE) method, based on a neural network with two outputs to build the endpoints of the PI.

Section II introduces the proposed approach to compute deterministic wind power forecasts and PIs. Section III presents a case study with data from Ireland. Section IV draws conclusions from this paper.

Floating Offshore Energy Devices   Materials Research Forum LLC
Materials Research Proceedings **20** (2022) 58-65   https://doi.org/10.21741/9781644901731-8

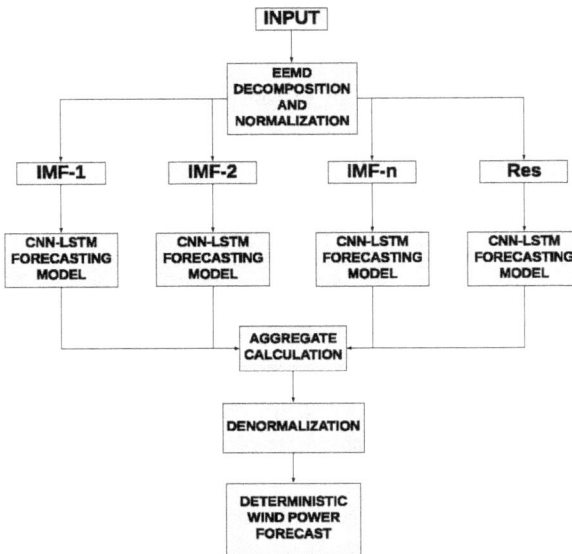

Fig. 1. Flowchart of the proposed model.

Methodology.

Deterministic forecasts. Empirical mode decomposition (EMD) [11] is a technique suitable for processing non-linear and non-stationary time series by dividing a series into modes known as intrinsic mode functions (IMFs). Nonetheless, EMD is susceptible to mode mixing. This issue can be overcome using EEMD [12], an enhanced version of EMD that adds different Gaussian white noise series of finite amplitude to the original signal. The various noise-added copies of the original signal are decomposed, and the mean value of the IMFs is taken as the result. This process helps to mitigate mode mixing.

Once the signal has been decomposed and normalized for every IMF, each one is trained separately by means of a CNN-LSTM architecture (Convolutional neural network and Long short-term memory). This neural network architecture allows to first extract features on input data with the CNN layer, while the LSTM layer [13] is a special recurrent neural network (RNN) structure that behaves as a memory cell and overcomes exploding and vanishing gradient problems that can occur with regular RNNs.

After each IMF has been trained, the original signal is denormalized and reconstructed to provide deterministic forecasts (Fig. 1).

In order to measure the performance of the proposed approach, the resulting point predictions are evaluated with the following metrics [14]: the normalized mean absolute error (NMAE), and the normalized root-mean-square error (NRMSE).

Floating Offshore Energy Devices

Materials Research Forum LLC

Materials Research Proceedings 20 (2022) 58-65

https://doi.org/10.21741/9781644901731-8

$$NMAE = \frac{1}{P_{inst} \cdot N} \sum_{i=1}^{N} |p_i - r_i|. \tag{1}$$

$$NMRSE = \frac{1}{P_{inst}} \sqrt{\frac{1}{N} \sum_{i=1}^{N} |p_i - r_i|}. \tag{2}$$

where N is the number of samples, p_i is the forecast wind power, r_i the actual wind power and P_{inst} is the total power capacity installed.

A CNN-LSTM architecture where wind power is not decomposed by EEDM is used as a benchmark to verify the accuracy of the model.

Prediction intervals. Deterministic forecasts cannot estimate the uncertainty of a given prediction. Therefore, the use of probabilistic forecasts is essential to obtain better economic results in the day-ahead electricity market [15].

PIs can be modelled by QR. The main advantage of this approach is that assumptions of any specific distribution are not needed. This methodology has been discussed in [16-18]. To obtain the forecast quantiles, simple ANN structures (CNN-LSTM, CNN and LSTM respectively) will be trained using the quantile regression loss function.

The reliability of the PI will be verified with an index termed as average coverage function (ACE) [19]:

$$ACE = PICP - PINC. \tag{3}$$

where PINC is the PI nominal coverage and PICP is the PI coverage probability, which is defined by:

$$PICP = \frac{1}{N} \sum_{i=1}^{N} c_i. \tag{4}$$

where N is the number of samples and c_i is a variable that indicates whether the measured value falls within the interval or not:

$$c_i = \begin{cases} 1, \text{if } r_i \in L_\alpha \\ 0, \text{if } r_i \notin L_\alpha \end{cases} \tag{5}$$

where L_α is the prediction interval. The PI will be more reliable the closest the ACE is to zero.

Case study

Data. The proposed approach has been tested and benchmarked with data from Ireland, obtained from EirGrid [20]. The wind power generation data ranges from 30-03-2019 to 03-07- 2019. As only historical wind power generation data were available, wind power is the only variable used as a predictor to train the model.

The dataset was divided into training, validation and testing sets to perform the study of the forecast model (Fig. 2). 80% percent of the data were employed to train the model. Considering the amount of data available, it is a reasonable proportion to train the model. The last five days of the dataset were used as a benchmark to compare the predictions with the actual values of wind power output.

Floating Offshore Energy Devices
Materials Research Proceedings **20** (2022) 58-65

Materials Research Forum LLC
https://doi.org/10.21741/9781644901731-8

Results. Errors for deterministic predictions for 1-h, 6-h and 24-h ahead forecasts for the proposed approach and the benchmark model are shown in Table 1. It can be seen that both NMAE and NRMSE are similar for the 1-h ahead forecasts. However, the EEDM-CNN-LSTM model clearly outperforms the benchmark model for longer forecast horizons. The increase of performance is notably high for the 24-h ahead forecast: the proposed approach obtains NMAE and NRMSE of 7.034% and 9.161%, whereas the CNN-LSTM model produces NMAE and NRMSE of 16.78% and 19.547% respectively.

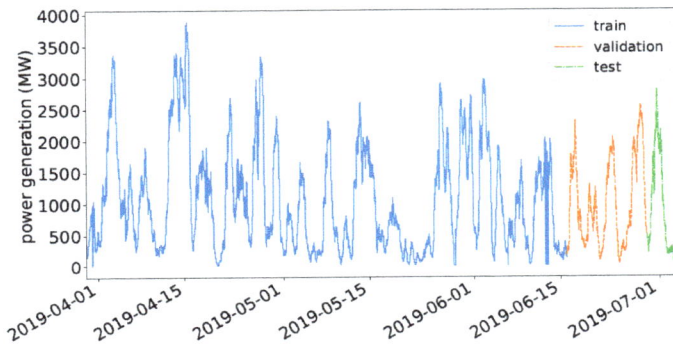

Fig. 2. Wind power time series.

Fig. 3 shows the prediction results for each horizon using both models. It is observed that the effect of decomposing the wind power into different sub-series using EEMD, and training each one separately, as it allows for the capture of the dynamics of the wind power output, such as spikes or sudden drops, in a more accurate way. On the other hand, the benchmark model can predict wind power fairly well 1-h ahead, but the accuracy of its forecasts decreases dramatically as the forecast horizon is increased.

The forecast quantiles were built by training three different models: a combined CNN-LSTM model, a simple CNN architecture, and a LSTM model. 95% and 80% PIs were computed 1-h and 6-h ahead. Table 2 shows the results of the different methods. Specifically, PICP and ACE have been used to evaluate the PIs. In terms of reliability, the CNN-LSTM model outperforms the other models in most of the scenarios. For instance, the 95% PIs constructed with the CNN-LSTM architecture for 1-h ahead forecasts produce a PICP of 91.25% in comparison to the PIs built by the CNN model (89.17%) and the LSTM model (83.75%). In this same scenario, the CNN-LSTM architecture has obtained a better ACE (-3.75%) than the ACE obtained by the other models (-5.83% and -11.25% for the CNN and LSTM models respectively).

The 95% and 80% PIs for every model are shown in Fig. 4. It can be observed that satisfactory PIs can be constructed by QR for short-term horizons.

Both proposed approaches to obtain deterministic predictions and PIs provide fairly reasonable results. These results could be improved by including more predictors, such as wind speed or numerical weather prediction data, and by training the model with a larger quantity of data, as limited amount of data has a negative impact on the performance of ANN models.

Floating Offshore Energy Devices
Materials Research Proceedings **20** (2022) 58-65

Materials Research Forum LLC
https://doi.org/10.21741/9781644901731-8

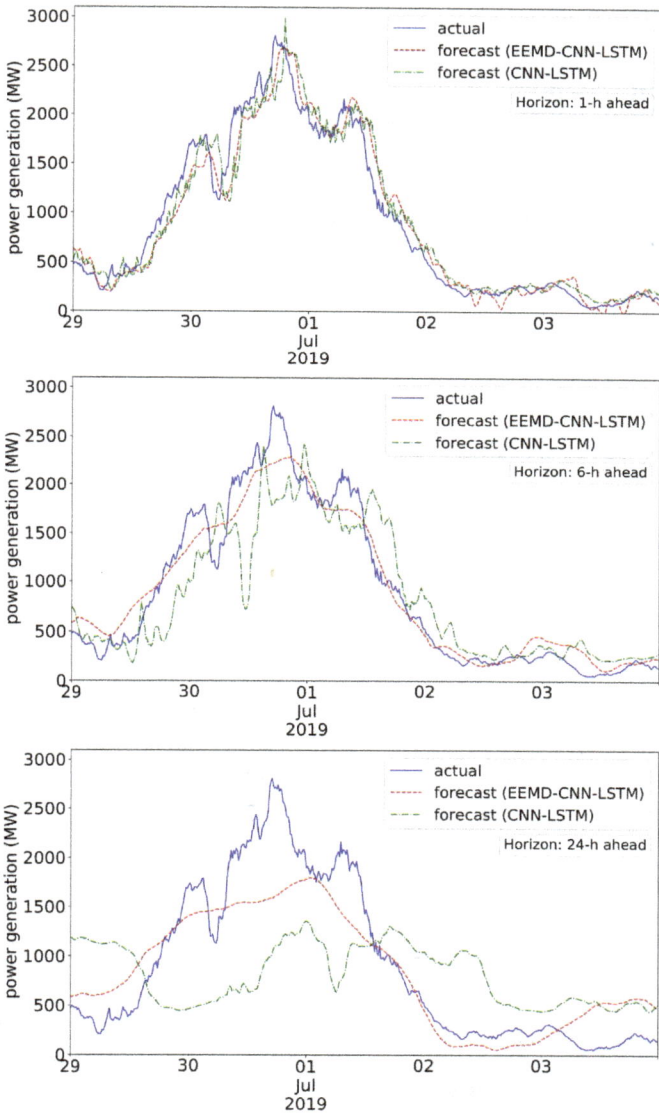

Fig. 3. Comparison of actual wind power and deterministic forecasts (1-h, 6-h, and 24-h ahead).

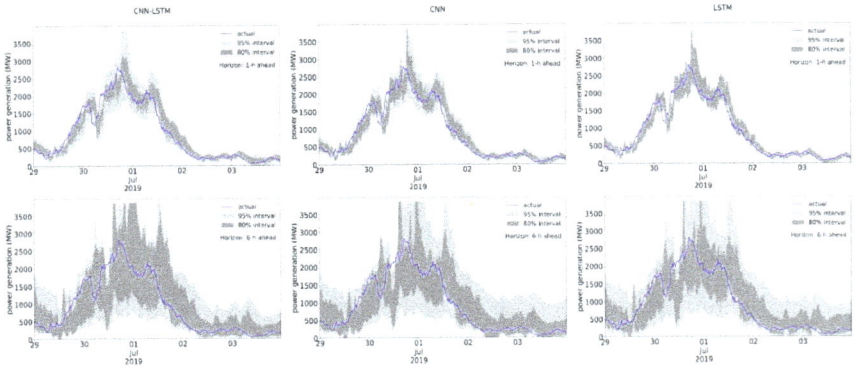

Fig. 4. Prediction intervals for 1-h and 6-ahead forecasts.

Table 1. Errors for deterministic forecasts.

Horizon	Error	CNN-LSTM	EEDM-CNN-LSTM
1-h	NMAE (%)	3.542	3.311
	NRMSE (%)	5.047	4.389
6-h	NMAE (%)	6.903	3.564
	NRMSE (%)	9.563	4.527
24-h	NMAE (%)	16.780	7.034
	NRMSE (%)	19.547	9.161

Table 2. Errors for prediction intervals.

95 (%)	CNN-LSTM		CNN		LSTM	
	PICP (%)	ACE (%)	PICP (%)	ACE (%)	PICP (%)	ACE (%)
1-h	91.25	-3.75	89.17	-5.83	83.75	-11.25
6-h	98.12	3.12	96.46	1.46	98.75	3.75
80 (%)	CNN-LSTM		CNN		LSTM	
	PICP (%)	ACE (%)	PICP (%)	ACE (%)	PICP (%)	ACE (%)
1-h	59.17	-20.83	64.17	-15.83	66.25	-13.75
6-h	88.33	8.33	83.75	3.75	78.75	-1.25

Conclusions

This paper presents an algorithm that combines EEDM and deep learning to provide short-term wind power forecasts, and the construction of PIs by QR. A case study using wind power generation data from Ireland has demonstrated that the proposed approach outperforms the benchmark model, especially for longer forecast horizons.

The model can improve in several ways. First, deep learning model benefits from having a larger quantity of data, and the data available in this case study were scarce. Therefore, increasing

Floating Offshore Energy Devices   Materials Research Forum LLC
Materials Research Proceedings **20** (2022) 58-65   https://doi.org/10.21741/9781644901731-8

their amount will result in better predictions. Secondly, only historical wind power data have been used as predictor. Multivariate models that include other variables such as wind speed or wind direction as predictors improve the final forecast as well. Finally, the whole installed power capacity was considered as a single wind farm, since data were not available on a wind farm level.

Afterwards, PIs were constructed by using QR. Three different models were used to obtain the forecast quantiles: a combined CNN-LSTM architecture, a CNN model, and a LSTM model. The reliability of the PIs was evaluated by using two indices: PICP and ACE. In most scenarios, the CNN-LSTM model provides the most reliable PIs. Further evaluation of the PIs can be done by assessing other parameters of the PI such as its sharpness [21]. As discussed for the deterministic model, forecast quantiles would benefit from larger amount of data and using other variables as predictors.

Acknowledgments
The authors acknowledge the funding of SEAI WindPearl Project 18/RDD/263. Vikram Pakrashi would like to acknowledge the support of SFI MaREI centre and UCD Energy Institute.

References
[1] M. A. Matos, R. J. Bessa, Setting the operating reserve using probabilistic wind power forecasts, IEEE Trans. Power Syst. 26 (2) (2010) 594-603. https://doi.org/10.1109/TPWRS.2010.2065818

[2] G. P. Swinand, A. O'Mahoney, Estimating the impact of wind generation and wind forecast errors on energy prices and costs in Ireland, Renew. Energy 75 (2015) 468-473. https://doi.org/10.1016/j.renene.2014.09.060

[3] G. Giebel, G. Kariniotakis, Wind power forecasting — a review of the state of the art, in: Renewable Energy Forecasting, Elsevier, 2017, pp. 59-109. https://doi.org/10.1016/B978-0-08-100504-0.00003-2

[4] H. Liu, H.Q. Tian, C. Chen, Y.F. Li, A hybrid statistical method to predict wind speed and wind power, Renew. Energy 35 (8) (2010) 1857-1861. https://doi.org/10.1016/j.renene.2009.12.011

[5] Y. Zhang, K. Liu, L. Qin, X. An, Deterministic and probabilistic interval prediction for short-term wind power generation based on variational mode decomposition and machine learning methods, Energy Convers. Manag. 112 (2016) 208-219. https://doi.org/10.1016/j.enconman.2016.01.023

[6] A. A. Abdoos, A new intelligent method based on combination of VMD and ELM for short term wind power forecasting, Neurocomputing 203 (2016) 111-120. https://doi.org/10.1016/j.neucom.2016.03.054

[7] J. B. Bremnes, Probabilistic wind power forecasts using local quantile regression, Wind Energy: An International Journal for Progress and Applications in Wind Power Conversion Technology 7 (1) (2004) 47-54. https://doi.org/10.1002/we.107

[8] R. J. Bessa, V. Miranda, A. Botterud, J. Wang, E. M. Constantinescu, Time adaptive conditional kernel density estimation for wind power forecasting, IEEE Trans. Sustainable Energy 3 (4) (2012) 660-669. https://doi.org/10.1109/TSTE.2012.2200302

Materials Research Forum LLC
https://doi.org/10.21741/9781644901731-8

[9] E. Xydas, M. Qadrdan, C. Marmaras, L. Cipcigan, N. Jenkins, H. Ameli, Probabilistic wind power forecasting and its application in the scheduling of gas-fired generators, Appl. Energy 192 (2017) 382-394. https://doi.org/10.1016/j.apenergy.2016.10.019

[10] A. Kavousi-Fard, A. Khosravi, S. Nahavandi, A new fuzzy-based combined prediction interval for wind power forecasting. IEEE Trans. Power Syst. 31 (1) (2015) 18-26. https://doi.org/10.1109/TPWRS.2015.2393880

[11] N. E. Huang, Z. Shen, S. R. Long, M. C. Wu, H. H. Shih, Q. Zheng, N.C. Yen, C. C. Tung, H. H. Liu, The empirical mode decomposition and the Hilbert spectrum for nonlinear and non-stationary time series analysis, Proceedings of the Royal Society of London. Series A: Mathematical, Physical and Engineering Sciences 454 (1971) (1998) 903-995. https://doi.org/10.1098/rspa.1998.0193

[12] Z. Wu, N. E. Huang, Ensemble empirical mode decomposition: a noise-assisted data analysis method, Adv. Adapt. Data Anal. (01) (2009) 1-41. https://doi.org/10.1142/S1793536909000047

[13] S. Hochreiter, J. Schmidhuber, Long short-term memory, Neural Comput. 9 (8) (1997) 1735-1780. https://doi.org/10.1162/neco.1997.9.8.1735

[14] H. Madsen, P. Pinson, G. Kariniotakis, H. A. Nielsen, T. S. Nielsen, Standardizing the performance evaluation of short-term wind power prediction models, Wind Engineering 29 (6) (2005) 475-489. https://doi.org/10.1260/030952405776234599

[15] J. Juban, N. Siebert, G. N. Kariniotakis, Probabilistic short-term wind power forecasting for the optimal management of wind generation, in: 2007 IEEE Lausanne Power Tech, IEEE, 2007, pp. 683-688. https://doi.org/10.1109/PCT.2007.4538398

[16] H. A. Nielsen, H. Madsen, T. S. Nielsen, Using quantile regression to extend an existing wind power forecasting system with probabilistic forecasts, Wind Energy: An International Journal for Progress and Applications in Wind Power Conversion Technology 9 (1-2) (2006) 95-108. https://doi.org/10.1002/we.180

[17] A. U. Haque, M. H. Nehrir, P. Mandal, A hybrid intelligent model for deterministic and quantile regression approach for probabilistic wind power forecasting, IEEE Trans. Power Syst. 29 (4) (2014) 1663-1672. https://doi.org/10.1109/TPWRS.2014.2299801

[18] C. Wan, J. Lin, J. Wang, Y. Song, Z. Y. Dong, Direct quantile regression for non-parametric probabilistic forecasting of wind power generation, IEEE Trans. Power Syst. 32 (4) (2016) 2767-2778. https://doi.org/10.1109/TPWRS.2016.2625101

[19] C. Wan, Z. Xu, P. Pinson, Z. Y. Dong, K. P. Wong, Probabilistic forecasting of wind power generation using extreme learning machine, IEEE Trans. Power Syst. 29 (3) (2013) 1033-1044. https://doi.org/10.1109/TPWRS.2013.2287871

[20] Wind Power Generation Data, http://smartgriddashboard.eirgrid.com/#all/wind, [Online, accessed 15-Jul-2019]

[21] P. Pinson, G. Kariniotakis, Conditional prediction intervals of wind power generation, IEEE Trans. Power. Syst. 25 (4) (2010) 1845-1856. https://doi.org/10.1109/TPWRS.2010.2045774

Floating Offshore Energy Devices
Materials Research Proceedings 20 (2022) 66-73

Materials Research Forum LLC
https://doi.org/10.21741/9781644901731-9

Frequency-Domain Identification of Radiation Forces for Floating Wind Turbines by Moment-Matching

Yerai Peña-Sanchez[1*], Nicolás Faedo[1] and John V. Ringwood[1]

[1] Centre for Ocean Energy Research, Maynooth University, Maynooth, Co. Kildare, Ireland

*e-mail: yerai.pena.2017@mumail.ie

Keywords: Radiation Convolution Term, Offshore Wind Turbine, Frequency-Domain Identification, Moment-Matching

Abstract. The dynamics of a floating structure can be expressed in terms of Cummins' equation, which is an integro-differential equation of the convolution class. In particular, this convolution operator accounts for radiation forces acting on the structure. Considering that the mere existence of this operator is highly inconvenient due to its excessive computational cost, it is commonly replaced by an approximating parametric model. Recently, the Finite Order Approximation by Moment-Matching (FOAMM) toolbox has been developed within the wave energy literature, allowing for an efficient parameterisation of this radiation force convolution term, in terms of a state-space representation. Unlike other parameterisation strategies, FOAMM is based on an interpolation approach, where the user can select a set of interpolation frequencies where the steady-state response of the obtained parametric representation exactly matches the behaviour of the target system. This paper illustrates the application of FOAMM to a UMaine semi-submersible-like floating structure.

Introduction

With the rapid decrease of the easily accessible fossil fuels, the immediate shift to renewable energy systems is one of the most important challenges of the 21st century. For this reason, the installed power capacity of renewable energy plants has significantly increased in this century, more than doubling it between 2004 and 2016 [1]. Among the renewable energy sources, wind energy has one of the highest growth, producing 4.4% of the worldwide electric power usage in 2017, and 11.6% electricity in the European Union [2]. In fact, onshore wind (and solar PV) will offer, in many places, a less expensive source of new electricity than the fossil-fuel alternative without financial assistance [3].

However, onshore wind has some disadvantages with respect to offshore wind farms. On the one hand, onshore wind farms have an impact on the landscape, since they usually require to be spread over more land than other conventional power stations, and need to be built in wild and rural areas, which can lead to habitat loss. On the contrary, offshore wind is steadier and stronger than onshore, with less visual impact. However, construction and maintenance costs are higher offshore.

At the moment, most offshore wind turbines are installed in shallow water (about $30m$ deep), using bot tom-mounted substructures. Nevertheless, to harness the available offshore wind potential, wind farms have to be located in deeper water. To this end, and to reduce the costs related with the structure, floating support-platforms will have to be deployed to hold the turbines, for which several platform configurations can be found in the literature [4, 5].

At this stage of development, simulations that combine aerodynamic, hydrodynamic and mooring-system dynamic effect on floating wind turbines are crucial to address the possible failure conditions of such structures and, therefore, accurately modelling each of those dynamical effects

Floating Offshore Energy Devices   Materials Research Forum LLC
Materials Research Proceedings 20 (2022) 66-73   https://doi.org/10.21741/9781644901731-9

is of paramount importance. In particular, the equation to describe the motion of a floating body, i.e. the equation describing the hydrodynamic interactions between the floating structure and the waves (the so-called Cummins' equation [6]), is an integro-differential equation, more precisely of the convolution class. The presence of this convolution operator represents a drawback for a number of reasons, including the fact that its numerical computation is highly inefficient, requiring considerable computational effort. To avoid such drawbacks, such a convolution operator can be approximated using a suitable parametric model (often given in terms of a state-space representation), for which several applications can be found, particularly within the wave energy literature.

In 2018, the Centre for Ocean Energy Research (COER) presented an identification strategy to compute a parametric model of the radiation convolution term of Cummins' equation [7], or the complete force-to-motion dynamics of a floating body. Such parameterisation strategy is based on recent advances in model order reduction by moment-matching, developed over several studies as, for example, [8]. The approach presented in [7] identifies a state-space model, whose frequency response exactly matches the frequency response of the target system at a set of user-selected frequencies. In fact, as a consequence of this interpolation feature, this moment-based strategy inherently preserves some of the relevant physical properties of the target floating body, such as internal stability and passivity. Motivated by the advantages behind moment-matching theory, reported in, for example, [7,9,10], a Matlab toolbox has been developed, to disseminate this moment-based identification strategy for wave energy applications [11].

The aim of the present paper is to introduce how FOAMM (Finite-Order Approximation by Moment-Matching) can be applied to compute parametric models of support platforms for offshore wind turbines, considerably reducing the computational effort related with time-domain simulation of floating structures. To illustrate the capabilities of such a toolbox, the *UMaine* semi-submersible-like floating structure [4,12] has been chosen as an application study, since its rigidly-connected multibodies represent a geometrically complex device, with frequency-response as shown in Section.

The remainder of this paper is organised as follows. Section 2 briefly introduces the equation of motion of a floating body, while the theory behind FOAMM is recalled in Section 3. Finally, an application case involving the *UMaine* structure is addressed in Section 3, whilst conclusions are encompassed in Section 4.

Equation of motion

Without any loss of generality, a single Degree of Freedom (DoF) support-platform is considered in this study. Recall that the motion of a 1-DoF support-platform can be expressed, in the time-domain, according to Newton's second law, obtaining the following linear hydrodynamic formulation [5]:

$$m\ddot{x}(t) = \mathcal{F}_r(t) + \mathcal{F}_{\hbar}(t) + \mathcal{F}_e(t) + \mathcal{F}_m(t), \tag{1}$$

where m is the mass of the structure (platform and turbine), $\ddot{x}(t)$ the acceleration of the body, $\mathcal{F}_e(t)$ the wave excitation force, $\mathcal{F}_r(t)$ the radiation force, $\mathcal{F}_{\hbar}(t)$ the hydrostatic restoring force, and $\mathcal{F}_m(t)$ the force exerted by the mooring system. The linearised hydrostatic force is given by $\mathcal{F}_{\hbar}(t) = -s_h x(t)$, where s_h denotes the hydrostatic stiffness. The mooring force is defined as $\mathcal{F}_m(t) = -b_m \dot{x}(t) - s_m x(t)$ [5], where b_m and s_m denote the damping and stiffness of the mooring system, respectively. From linear potential theory, $\mathcal{F}_r(t)$ can be modelled using Cummins' equation [6] as,

Floating Offshore Energy Devices   Materials Research Forum LLC
Materials Research Proceedings 20 (2022) 66-73   https://doi.org/10.21741/9781644901731-9

$$\mathcal{F}_r(t) = -\mu_\infty \ddot{x}(t) - \int_{\mathbb{R}^+} k(\tau)\dot{x}(t-\tau)d\tau, \tag{2}$$

where $\mu_\infty = \lim_{\omega \to +\infty} A(\omega) > 0$ denotes the radiation added-mass at infinite frequency, and $k(t)$ $\in L^2(\mathbb{R})$ is the radiation impulse response function. Eq. (1) can be rewritten as:

$$(m + \mu_\infty)\ddot{x}(t) + k(t) * \dot{x}(t) + s_h x(t) = \mathcal{F}_e(t) + \mathcal{F}_m(t), \tag{3}$$

where the symbol $*$ represents the convolution operator.

Since this paper is focused on the approximation of the radiation convolution term, and FOAMM identifies a parametric form using raw frequency-domain data, it is convenient to define the frequency-domain equivalent of the radiation convolution term, which can be obtained through Ogilvie's relations [15] as:

$$B(\omega) = \int_{\mathbb{R}^+} k(t) \cos(\omega t)\, dt, \quad A(\omega) = \mu_\infty - \frac{1}{\omega}\int_{\mathbb{R}^+} k(t)\sin(\omega t)\, dt \tag{4}$$

where the coefficients $B(\omega)$ and $A(\omega)$ represent the radiation damping and added-mass of the device, respectively. This set of hydrodynamic coefficients and can be efficiently obtained using any of the state-of-the-art Boundary Element Method (BEM) solvers (see [16]). The impulse response function $k : \mathbb{R}^+ \to \mathbb{R}$ can be written as a mapping involving the radiation damping coefficient as:

$$k(t) = \frac{2}{\pi}\int_{\mathbb{R}^+} B(\omega)\cos(\omega t)\, d\omega. \tag{5}$$

with frequency-domain equivalent given by

$$K(\omega) = B(\omega) + j\omega\, [A(\omega) - \mu_\infty]. \tag{6}$$

Moment-matching-based parameterization
To keep this paper reasonably self-contained, this section provides a brief summary of the theory behind FOAMM. The interested reader is referred to [7] for an extensive discussion on the specific underlying mathematical principles.

The radiation impulse response mapping defines a linear-time invariant system completely characterised by $k(t)$, where its input is the body velocity, i.e. $\dot{x}(t)$. To be precise, the radiation subsystem Σ^k is given

$$\Sigma_k : \theta_k(t) = k(t) * \dot{x}(t), \tag{7}$$

where $\theta_K(t) \in \mathbb{R}$ is the output (radiation force) of system Σ_k.

To obtain a parametric description of (7), the velocity of the floating structure $\dot{x}(t)$ is expressed as an autonomous signal generator as,

$$\mathcal{G}_{\dot{x}} : \{\dot{\xi}_{\dot{x}}(t) = S\xi_{\dot{x}}(t), \quad \dot{x}(t) = L_{\dot{x}}\xi_{\dot{x}}(t) \tag{8}$$

where the matrix S is defined [7] as

$$S = \oplus_{p=1}^{\beta} \begin{bmatrix} 0 & \omega_p \\ -\omega_p & 0 \end{bmatrix} \tag{9}$$

where the symbol $\oplus$ denotes the direct sum of matrices of β matrices, and $\nu = 2\beta$, with β the number of interpolation frequencies. Note that each $\omega_p \in \mathcal{F}$, with $\mathcal{F} = \{\omega_i\}_{i=1}^{\beta} \subset \mathbb{R}^+$ represents a desired interpolation point for the moment-matching-based parameterisation process, i.e. a frequency where the transfer function of the parametric model matches the transfer function of the target system.

Following [7], the so-called moment-domain equivalent of the output of system Σ_k in (7) can be straightforwardly computed as

$$\underline{\mathcal{Y}}_k = L_{\dot{x}} \mathcal{R}^k, \tag{10}$$

where the matrix $\mathcal{R}^k$ is defined by

$$\mathcal{R}^k = \oplus_{p=1}^{f} \begin{bmatrix} \Re\{K(j\omega_p)\} & \Im\{K(j\omega_p)\} \\ -\Im\{K(j\omega_p)\} & \Re\{K(j\omega_p)\} \end{bmatrix} \tag{11}$$

Finally, the parametric (state-space) description

$$\widetilde{\Sigma_{k\mathcal{F}}}: \{\dot{\Theta}_k(t) = F_k \Theta_k(t) + G_k \dot{x}(t), \quad \widetilde{\theta}_k(t) = Q_k \Theta_k(t) \tag{12}$$

is a system that interpolates the target frequency response $K(j\omega)$ at the set $\mathcal{F}$, i.e. it has the *exact* same frequency response of the radiation subsystem Σ_k at the frequencies defined in the set $\mathcal{F}$, if $Q_k P_k = \underline{\mathcal{Y}}_k$, where P_k is the unique solution of the Sylvester equation

$$F_k P_k + G_k L_{\dot{x}} = P_k S, \tag{13}$$

and $\underline{\mathcal{Y}}_k$ is computed from equation (10). The reader is referred to [7] for the theory behind the explicit computation of the matrices F_k, G_k, Q_k in (12) fulfilling condition (13).

Application example

The selected support-platform is a *UMaine* semi-submersible-like floating structure, constrained to move in pitch. This structure is designed to support the multi-megawatt turbine *NREL offshore 5-MW baseline wind turbine* [17]. In this study, this structure is selected due to its complex geometry (illustrated in Figure 1). In other words, the radiation convolution frequency-response for this device is more geometrically complex than for other floating bodies analysed before, such as, for example, in [7,11,9]. More information about the specifications of the *Umaine* structure is provided in 1.

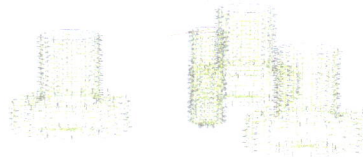

Figure 1: Low-order mesh of the UMaine semi submersible-like structure analysed in this study.

Floating Offshore Energy Devices
Materials Research Proceedings 20 (2022) 66-73

Materials Research Forum LLC
https://doi.org/10.21741/9781644901731-9

Table 1: Specifications of the structure.

Column spacing	50m
Main column diameter	6.5m
Side columns diameter	12 and 24m
Draft	20m
Mass (with ballast)	$13.5 \cdot 10^6$
Center of mass (z)	-13.74 from SWL

Figure 2 shows the radiation damping and added-mass in the top and bottom left-hand side figures, respectively, along with the frequency-response of the convolution operator, $K(j\omega)$, in the right-hand-side of the figure (both magnitude and phase) computed as shown in 6.

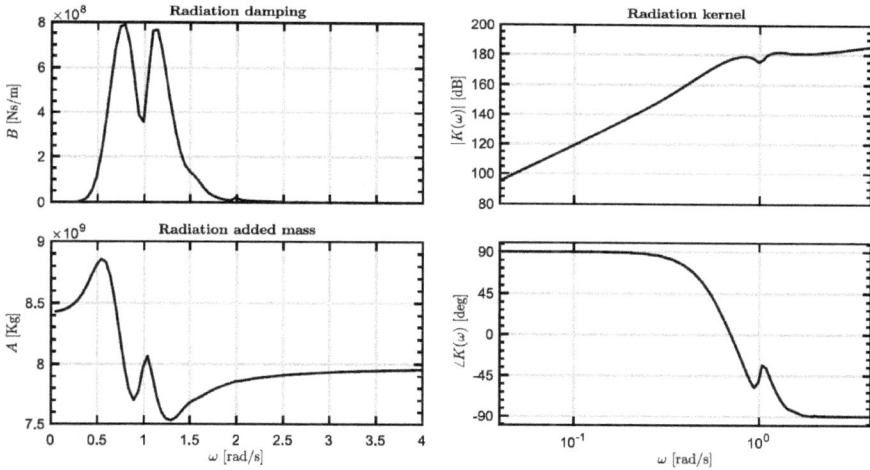

Figure 2: Hydrodynamic coefficients for the pitch motion of the UMaine semi-submersible like device analysed in this Study. The radiation damping and added-mass are represented in the left-hand-side of the figure, while the radiation force frequency-response is illustrated on the right-hand-side.

To compute the parametric model of the radiation convolution term shown in Figure 2, the software FOAMM needs to be downloaded first, which can be done for free from http://www.eeng.nuim.ie/coer/downloads/. Finally, as reported in [11], it is necessary to install the correct Matlab runtime version, which is readily provided with FOAMM. The interested reader is referred to [11] for more information on the different options and modes available on this toolbox.

The first choice for the user is the frequency range over which the parameterisation is carried out. As explained in [7], such a frequency range highly depends on the application, and it is usually conditioned by the typical sea-state characterising the location of the structure. Since this is not relevant for the current study, the frequency range is selected as $\omega_l = 0.3$ [rad/s] and $\omega_u = 3$ [rad/s].

Floating Offshore Energy Devices

Materials Research Forum LLC

Materials Research Proceedings **20** (2022) 66-73

https://doi.org/10.21741/9781644901731-9

For the selection of the interpolation frequencies, three different methods are available in FOAMM [11]. This study utilises the so-called manual identification method, where the user selects the set of interpolation frequencies. This presents several advantages since, by selecting the interpolation frequencies in a sensible manner (dynamically speaking), the accuracy of the parametric model can be considerably improved. Due to the complexity of the *UMaine* frequency-response, 7 frequencies (parametric model with order 14) are required to obtain an accept approximation, with a Mean Absolute Percentage Error (MAPE) of ≈ 0.08 %.

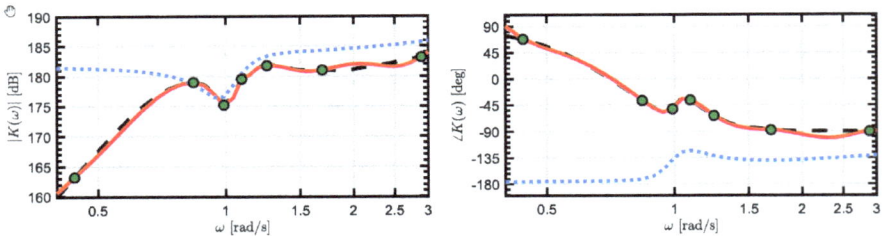

Figure 3: Frequency-response of the parametric model obtained with FOAMM for the radiation force subsystem (solid-red), along with the (target) frequency-response of the Umaine structure (dashed-black), the set of interpolation frequencies considered (green dots), and the best approximated model using subspace-methods (dotted-blue).

To provide a comparison with well-established identification strategies, a parameterisation using subspace-methods (implemented in the **Matlab** function *n4sid*) is considered. The most accurate model obtained with this method has a MAPE of 0.45 % of error. The frequency-response of the obtained parametric model obtained using FOAMM, along with the target frequency-domain data of the *Umaine* structure, the set of interpolation frequencies considered, and the best approximated model using subspace-methods, are shown in Figure 3.

Conclusions

This paper illustrates how to use the FOAMM toolbox to obtain a parametric model of the convolution term associated to radiation forces, for a complex-shape support- structure of a floating wind turbine. The chosen floating structure is the *UMaine* semi- submersible-like device which, due to its geometrical complexity, represents a challenge from a frequency-domain identification perspective. In fact, it is shown that, while no accurate approximation model could be obtained using well-established subspace-methods, FOAMM provides an accurate parametric description of the radiation force subsystem as a consequence of its interpolation features, which allows the user for the selection of the frequencies characterising the dynamics of the structure as interpolation points in the parameterisation process.

Acknowledgment

This material is based upon works supported by the Science Foundation Ireland under Grant No. 13/IA/1886.

References

[1] REN21, "Renewables 2017: Global status report," REN21 Renewables now, Tech. Rep., 2017.

[2] W. Europe, "Wind in power: 2017 european statistics," *Bruksela: Wind Europe*, 2017.

[3] IRENA, "Renewable energy and jobs: Annual review 2019," International Renewable Energy Agency, Tech. Rep., 2019.

[4] A. N. Robertson, J. M. Jonkman *et al.*, "Loads analysis of several offshore floating wind turbine concepts," in *The Twenty-first International Offshore and Polar Engineering Conference*. International Society of Offshore and Polar Engineers, 2011.

[5] Y.-H. Lin, S.-H. Kao, and C.-H. Yang, "Investigation of hydrodynamic forces for floating offshore wind turbines on spar buoys and tension leg platforms with the mooring systems in waves," *Applied Sciences*, vol. 9, no. 3, p. 608, 2019. https://doi.org/10.3390/app9030608

[6] W. Cummins, "The impulse response function and ship motions," DTIC Document, Tech. Rep., 1962.

[7] N. Faedo, Y. Peña-Sanchez, and J. V. Ringwood, "Finite-order hydrodynamic model determination for wave energy applications using moment-matching," *Ocean Engi- neering*, vol. 163, pp. 251–263, 2018. https://doi.org/10.1016/j.oceaneng.2018.05.037

[8] A. Astolfi, "Model reduction by moment matching for linear and nonlinear systems," *IEEE Transactions on Automatic Control*, vol. 55, no. 10, pp. 2321–2336, 2010. https://doi.org/10.1109/TAC.2010.2046044

[9] N. Faedo, Y. Peña-Sanchez, and J. V. Ringwood, "Passivity preserving moment- based finite-order hydrodynamic model identification for wave energy applications," in *Proceedings of the 3rd International Conference on Renewable Energies Offshore, Lisbon*, 2018. https://doi.org/10.1016/j.oceaneng.2018.05.037

[10] Y. Peña-Sanchez, N. Faedo, and J. V. Ringwood, "Moment-based parametric identi- fication of arrays of wave energy converters," in *Submitted to 2019 American Control Conference, Philadelphia*, 2019.

[11] Y. Peña-Sanchez, N. Faedo, M. Penalba, G. Giorgi, A. Mérigaud, C. Windt, D. García Violini, L. Wang, and J. V. Ringwood, "Finite-Order hydrodynamic Ap- proximation by Moment-Matching (FOAMM) toolbox for wave energy applications," in *Proceedings of the 13th European Wave and Tidal Energy Conference, EWTEC, Naples, Italy*, 2019.

[12] M.J. Coulling, A. J. Goupee, A. N. Robertson, J. M. Jonkman, and H. J. Dagher, "Validation of a fast semi-submersible floating wind turbine numerical model with deepcwind test data," *Journal of Renewable and Sustainable Energy*, vol. 5, no. 2, p. 023116, 2013. https://doi.org/10.1063/1.4796197

[13] E. Kristiansen, Å. Hjulstad, and O. Egeland, "State-space representation of radiation forces in time-domain vessel models," *Ocean Engineering*, vol. 32, no. 17, pp. 2195– 2216, 2005. https://doi.org/10.1016/j.oceaneng.2005.02.009

[14] T. Pérez and T. I. Fossen, "Time-vs. frequency-domain identification of parametric radiation force models for marine structures at zero speed," *Modeling, Identification and Control*, vol. 29, no. 1, pp. 1–19, 2008. https://doi.org/10.4173/mic.2008.1.1

[15] T. F. Ogilvie, "Recent progress toward the understanding and prediction of ship motions," in *Proceedings of the 5th ONR Symposium on Naval Hydrodynamics, Bergen*, 1964.

[16] J. Falnes, Ocean waves and oscillating systems: linear interactions including wave- energy extraction. Cambridge university press, 2002. https://doi.org/10.1017/CBO9780511754630

[17] J. Jonkman, S. Butterfield, W. Musial, and G. Scott, "Definition of a 5-mw ref- erence wind turbine for offshore system development," National Renewable Energy Lab.(NREL), Golden, CO (United States), Tech. Rep., 2009. https://doi.org/10.2172/947422

Floating Offshore Energy Devices
Materials Research Proceedings 20 (2022) 74-80

Materials Research Forum LLC
https://doi.org/10.21741/9781644901731-10

Design and Structural Testing of Blades for a 2MW Floating Tidal Energy Conversion Device

Yadong Jiang[1,2,3,a], Edward Fagan[1,2,3,b], William Finnegan[1,2,3,c],

Afrooz Kazemi Vanhari[1,2,3,d], Patrick Meier[1,2,3,e], Suhaib Salawdeh[3,f],

Colm Walsh[3,g] and Jamie Goggins[1,2,3,h,*]

[1]Centre for Marine and Renewable Energy Ireland (MaREI), Environmental Research Institute, Ringaskiddy, Co. Cork P43 C573, Ireland

[2]Ryan Institute, National University of Ireland Galway, University Road, Galway H91 HX31, Ireland

[3]School of Engineering, National University of Ireland Galway, University Road, Galway H91 HX31, Ireland

[a]yadong.jiang@nuigalway.ie, [b]edward.fagan@nuigalway.ie, [c]william.finnegan@nuigalway.ie, [d]afrooz.kazemi@nuigalway.ie, [e]patrick.meier@nuigalway.ie, [f]suhaib.salawdeh@nuigalway.ie, [g]walshc@nuigalway.ie, [h]jamie.goggins@nuigalway.ie

Keywords: Low Carbon Emissions, Manufacturing, Advanced Materials, Tidal Energy, Tidal Turbine

Abstract. The floating tidal energy is increasingly recognised to have the potential of delivering a step-change cost reduction to the tidal energy sector, by extracting energy from deeper water sites through energy conversion devices. To ensure the normal operation of a tidal energy convertor within its service life, the device should be designed properly and evaluated through a series of strength and durability testing. The Large Structures Research Group at NUI Galway is working closely with, renewable energy company, Orbital Marine Power and, blade manufacture, ÉireComposites Teo, to design and test the next generation of SR2000 tidal turbine blade, with aims to increase the turbine power production rate and to refine the design for low cost. This paper presents a brief description of the structural design and testing of a blade for the O2-2000 tidal turbine, one of the largest tidal turbines in the world. NUI Galway will utilise their in-house software, BladeComp, to find a blade laminates design that balances both blade strength and material cost. The structural performance of the designed blade will be assessed by conducting static and fatigue testing. To achieve this objective, a support frame to fix the blade is designed, a load application device is introduced and the methodology for design tidal loading conversion is proposed in order to complete the testing at NUI Galway.

Introduction

After many years of delay, tidal stream energy is now becoming a commercial reality. The MeyGen project is set to become the first 4-turbine, 6MW tidal array [1]. At the same time, EDF is committed to projects in France using the OpenHydro/DCNS tidal device and projects are also being considered in the Bay of Fundy in Canada and elsewhere [2]. A number of market assessments for tidal stream energy have been independently developed and, as with any early-stage technology sector, there is a wide degree of variation between projections. The International Energy Agency's 'Blue Map' [3] medium growth scenario predicts 13GW of installed tidal capacity by 2050, with a high growth scenario of 52GW over the same period.

Floating Offshore Energy Devices
Materials Research Proceedings **20** (2022) 74-80

Materials Research Forum LLC
https://doi.org/10.21741/9781644901731-10

Despite the huge growth potential in the tidal energy market, there is only limited data available about how composite materials will perform under high cyclic loading in harsh marine conditions. Moreover, there is no published work available for the full-scale fatigue testing of tidal blades. Therefore, the Large Structures Research Group, NUI Galway is collaborating with Orbital Marine Power (OMP) Limited in the H2020 FloTEC and the OCEANERA-COFUND SEABLADE projects to design and test a full-scale blade for the OMP tidal turbines.

The Large Structures Research Group at NUI Galway has many years of experience in structural design and processing of glass and carbon fibre-reinforced composite materials. As a member of the MaREI Centre, the group has developed advanced computational design methodologies [4,5] for tidal current turbine blades, performed design and optimization studies on wind turbine blade structures of several scales [6]-[8], and conducted structural testing of components for a 3/8th scale blade and rotor subsection for the OpenHydro prototype tidal turbine [9].

Orbital Marine Power (OMP) Limited is credited with pioneering floating tidal stream turbines since the company's formation in 2002 in Orkney, Scotland. OMP has maintained and advanced this position by developing the world's leading engineering knowledge and technology in floating tidal stream turbines. The OMP SR2000 produced unrivalled performance during a demonstration programme, where it delivered multiple world-firsts, including exporting over 3,250 MWh of electricity to the Orkney grid during a 12 month period. This was more than the entire wave and tidal sector in Scotland had exported over the 12 years prior to the launch of the SR2000.

In this paper, the two aspects detailed are the structural design and testing of the O2-2000 turbine blade, which is the next generation of the SR2000 blade. The NUI Galway in-house developed software BladeComp is utilised to design and optimise the layups of the O2-2000 blade in order to balance both blade strength and manufacturing costs. The structural testing aims to evaluate the blade performance under both extreme static loading and long-term fatigue loading, which will be conducted in the Large Structures Research Laboratory of NUI Galway.

Blade Design

The turbine blade targeted in this research is designed for the OMP floating tidal energy converter (Fig. 1). It has a capacity of 2 MW and is equipped with two 20 m diameter twin-bladed rotors, which make it one of the largest tidal turbine systems in the world. The aerodynamic design, which addressed the external shape of the O2-2000 blade (Fig. 2), was conducted by OMP. The structural design of the blade has been generated and assessed using BladeComp, which automates the process of generating, analysing and post-processing finite element analysis models of tidal turbine blades. This software acts as a wrapper for the finite-element (FE) software, utilising the advanced mechanic features of the FE technics and tailoring the analysis to specifically address the design of tidal turbine blades based on genetic algorithm.

The designed O2-2000 turbine blade consists of a single internal shear web. The blade trailing edge fairings and the blade tip are constructed separately to the main body and will be adhesively bonded to the finished structure. The O2-2000 turbine blades are constructed from glass-fibre epoxy, using a "semi-preg" powder epoxy material technology developed by project partners ÉireComposites Teo. The laminates that comprise the turbine elements are reinforced with unidirectional and biaxial E-glass plies. The biaxial plies are used in the leading and trailing edge sections of the blade shells and the shear webs. The blade sections subjected to bending stress (the spar caps and root region) will include significant unidirectional reinforcement. The flapwise and edgewise design loading profiles defined for the blade design and structural testing were supplied by OMP.

Floating Offshore Energy Devices
Materials Research Proceedings **20** (2022) 74-80

Materials Research Forum LLC
https://doi.org/10.21741/9781644901731-10

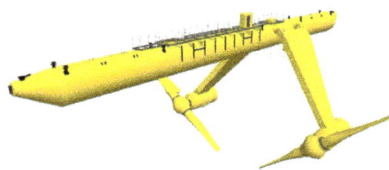

Fig. 1 The Orbital Marine Power Ltd designed floating tidal energy convertor

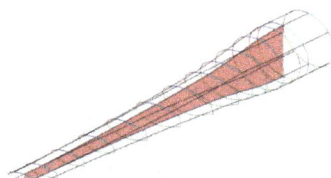

Fig. 2 The Orbital O2-2000 tidal turbine blade

Structural Testing

With the O2-2000 tidal turbine blade designed and manufactured, the structural testing, which includes both static and fatigue mechanical tests, will be conducted in the Large Structures Research Laboratory located in NUI Galway. Fig. 3 shows the test setup overview. The blade is supported at its root on the support frame and will be loaded via three hydraulic actuators ranging in capacities from 240 kN to 750 kN. The actuators can be controlled separately, and thus will enable the application of complex loading patterns in the static and fatigue testing. The load amplitudes of the actuators are converted from the design loading profiles to simulate the operating conditions of the blade underwater.

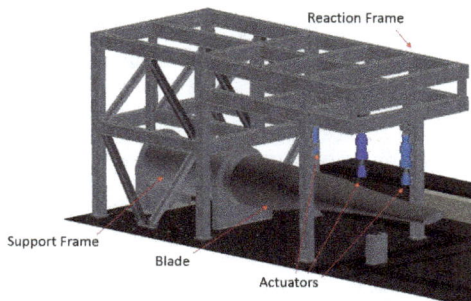

Fig. 3 Overview of the test setup

Fig. 4 The support frame

Support Frame. During the testing, the blade root is expected to be fully constrained. To achieve this, a root support frame was designed, which is shown in Fig. 4, where there is a ring of bolt holes drilled on the front surface of the support frame, with a pitch hole diameter of 1640 mm. However, the pitch hole diameter of the O2-2000 blade root is 1400 mm. Thus, an adopter plate, with two rings of bolt holes drilled on its surface, is designed to mount the blade root on the support frame. To prevent the blade root movements during the static and fatigue testing, the support frame is connected to the reinforced concrete reaction floor of the laboratory through pre-tensioned bolts.
Load Application. Three hydraulic actuators will be used in the static and fatigue testing for load application purpose. Swivel connections are used to mount the actuators to the reaction frame. Thus, no moments will be introduced to the blade. During operation, the whole blade external surface will suffer tidal loads. Thus, more load points will result in better simulation of blade operation conditions. Since there are only three actuators available, a load introduction device is designed

Floating Offshore Energy Devices   Materials Research Forum LLC
Materials Research Proceedings **20** (2022) 74-80   https://doi.org/10.21741/9781644901731-10

to introduce more point loads. As shown in Fig. 5, the triangle-shaped device can transfer a single load from actuator to two bottom contact surfaces. By employing this device, the load point number is doubled, which allows a good simulation of tidal loads. To spread the point loads uniformly to the blade surface and avoid the local damage, a clamp device is used, which is shown in Fig. 6 and is a steel section filled with plywood or Nylon. The clamp comprises of two parts, a pressure side and a suction side. The inner surface is in contact with the blade external surface. A 5 mm thick rubber layer will be padded on the blade surface to avoid any damages. The clamp allows for testing the blade both in the pressure-to-suction and suction-to-pressure directions without any change in the test set-up, which is capable of both static and fatigue testing.

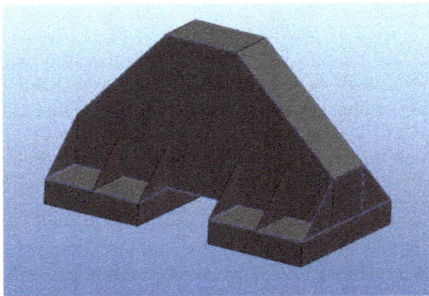

Fig. 5 The load introduction device

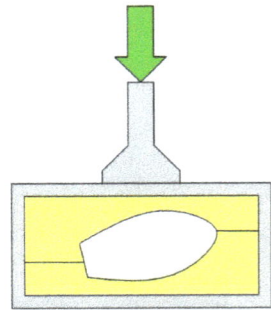

Fig. 6 The clamp for load introduction

Data Acquisition. The strain values on the blade surface will be measured by the electrical resistance strain gauges to monitor the damage occurrence of the composite materials. Two types of displacement transducers, namely the draw-wire string potentiometers and the LVDTs, will be employed to measure the deflection of the blade and the root movements of the blade root connection. Additionally, a Digital Image Correlation (DIC) and a laser scanner will be utilised to supply more detailed information about the blade deformation. The blade natural frequencies will be measured by a laser vibrometer to estimate the blade damage level under fatigue loading.

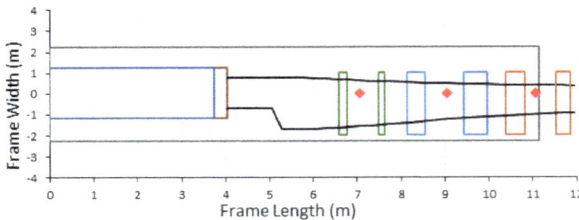

Fig. 7 The load application locations of the blade

Floating Offshore Energy Devices   Materials Research Forum LLC
Materials Research Proceedings 20 (2022) 74-80   https://doi.org/10.21741/9781644901731-10

Test Loading Definition. The tidal loads can be decomposed into flapwise and edgewise components. In the static testing, to simulate the extreme loading condition of the tidal turbine, the loads in two directions should be applied to the blade simultaneously. But due to the facility limitation, there are only three actuators available in the laboratory which only allows for applying flapwise loads. To overcome this issue, the blade will be installed with a specific pitch angle. By pitching the blade, an attack angle is introduced to the actuator load direction, which will introduce additional edgewise loading to the blade. It should be noted that the load point locations (Fig. 7) cannot be adjusted as the actuators are fixed. Thus, only the blade pitch angle and the actuator amplitudes can be tuned to make the blade load profiles (including moments and shear loads) under testing loading agree well with that under design loading. For this purpose, a genetic algorithm based calibration was conducted. The variables to address in the calibration are the blade pitch angle and the actuator load amplitudes. The goodness-of-fit is judged according to the mean error between the blade moment and shear profiles under the design and calibrated loading. The calibration results show that the optimised blade pitch angle should be 12.5°. The blade moment profiles under the design loading and the calibrated test loading are compared in Fig. 8. Good agreement can be found in these comparisons, which indicates that the method of pitching blade to get the flapwise and edgewise loads applied simultaneously is suitable for the static testing.

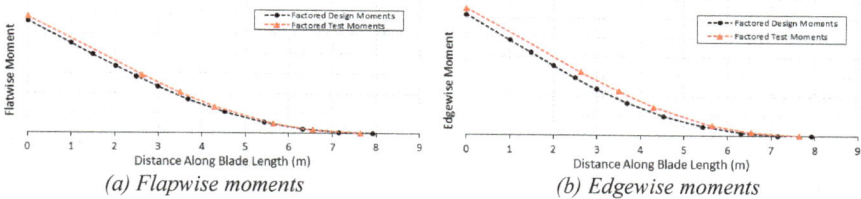

(a) Flapwise moments   *(b) Edgewise moments*

Fig. 8 Design loads and converted testing loads for the static test

Regarding the fatigue testing, the damage equivalent load case defined in the Blade Data Pack [10] is used. Similar to the static testing, calibration was conducted to obtain the blade pitch angle and actuator amplitudes. It was found that the best blade pitch angle for applying static loading is 15°. However, it is estimated that at least two weeks of additional time would be required to remove the instrumentation, disconnect and reinstall the rotated blade and reinstall the instrumentation. Thus, decision was made that the blade pitch angle used in fatigue testing keeps the same as that of static testing to reduce the testing period. Fig. 9 shows the blade load profile comparisons under design fatigue loading and calibrated fatigue loading. It could be found that the blade load profiles under calibrated fatigue loading agree well with that under design fatigue loading, without changing the blade pitch angle.

The maximum operating frequency of the actuator is dependent on its displacement amplitude. In fatigue testing, all three actuators will be adjusted to operate under the same frequency. The initial estimates for allowable actuator displacements of ±39 mm, ±18 mm and ±6 mm for actuators 3, 2 and 1, respectively. This will result in a loading frequency of 0.33 Hz, which leads to a test duration of approximately 35 days (of continuous operation) for 1,000,000 cycles. It should be noted that with consideration of stoppages for regular inspections of blade and test setup, the testing period is scheduled for approximately 2 months.

Floating Offshore Energy Devices
Materials Research Proceedings **20** (2022) 74-80

Materials Research Forum LLC
https://doi.org/10.21741/9781644901731-10

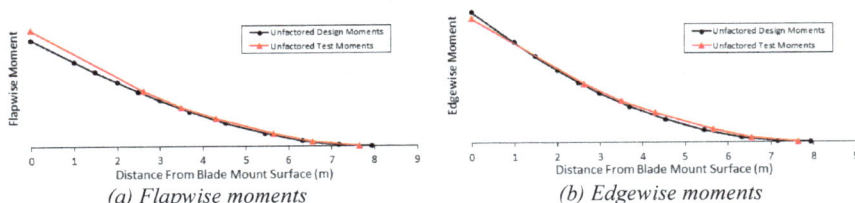

(a) Flapwise moments

(b) Edgewise moments

Fig. 9 Design loads and converted testing loads for the fatigue test

Summary and Future Work

This paper focuses on the design and testing of a blade for a 2MW floating tidal energy convertor which is developed by OMP. The design of the tidal blade is briefed with considerations of reducing the material costs and increasing the blade strength. The preparation of the blade static and fatigue testing is illustrated, including the support frame, the load introduction mechanism and the test loading definition. Currently, the designed tidal turbine blade is being manufactured. Based on the test specifications described in this paper, the strength and durability of the blade will be evaluated through the static and fatigue testing, respectively. The test results will not only provide valuable data to gain the confidence of OMP in the commercialisation of the O2-2000 tidal turbine but also contribute to the field of full-scale testing of tidal turbine blade.

Acknowledgement

The authors would like to acknowledge the Science Foundation Ireland (SFI) for funding this project through MaREI, the SFI Research Centre for Energy, Climate and Marine (grant no. 12/RC/2302), the European Union's Horizon 2020 research and innovation programme for funding the research through the FloTEC project (grant no. 691916) and under the OCEANERA-NET COFUND SEABLADE project (grant no. 731200). The last author would like to acknowledge the support of SFI through the Career Development Award programme (Grant No. 13/CDA/2200). Additional acknowledgements are given to the technical staff at NUI Galway and engineering staff at Orbital Marine Power Limited and ÉireComposites Teo.

References

[1] MeyGen project, https://simecatlantis.com/projects/meygen/

[2] Marine Current Energy: One Current to Another, https://www.edf.fr/en/the-edf-group/industrial-provider/renewable-energies/marine-energy/marine-current-power

[3] Energy Research Partnership, Scenario analysis, International Energy Agency, 'BLUE Map', 2010.

[4] C.R. Kennedy, S.B. Leen C.M. Ó Brádaigh, A Preliminary Design Methodology for Fatigue Life Prediction of Polymer Composites for Tidal Turbine Blades, Proceedings of the Institution of Mechanical Engineers, Part L, Journal of Materials: Design and Applications, 226 (2012), 203-218. https://doi.org/10.1177/1464420712443330

[5] E.M. Fagan, S.B. Leen, C.R. Kennedy, J. Goggins, Damage mechanics based design methodology for tidal current turbine composite blades, Renewable Energy, 97 (2016), 358-72. https://doi.org/10.1016/j.renene.2016.05.093

Materials Research Forum LLC
https://doi.org/10.21741/9781644901731-10

[6] E.M. Fagan, M. Flanagan, S.B. Leen, T. Flanagan, A. Doyle, J. Goggins, Physical experimental static testing and structural design optimisation for a composite wind turbine blade, Composite Structures, 164 (2016), 90-103. https://doi.org/10.1016/j.compstruct.2016.12.037

[7] E.M. Fagan, S.B. Leen, O. De La Torre, J. Goggins, Experimental investigation, numerical modelling and multi-objective optimisation of composite wind turbine blades, Journal of Structural Integrity and Maintenance, 2 (2017), 109-119. https://doi.org/10.1080/24705314.2017.1318043

[8] E.M. Fagan, O. De La Torre, S.B. Leen, J. Goggins, Validation of the multi-objective structural optimisation of a composite wind turbine blade, Composite Structures, 204 (2018), 567-577. https://doi.org/10.1016/j.compstruct.2018.07.114

[9] De La Torre, D. Moore, D. Gavigan, J. Goggins, Accelerated life testing study of a novel tidal turbine blade attachment, International Journal of Fatigue, 114 (2018), 226-237. https://doi.org/10.1016/j.ijfatigue.2018.05.029

[10] F.Wallace, SR2-2000 Blade Data Pack, Research Data, 2017.

Floating Offshore Energy Devices

Materials Research Proceedings **20** (2022) 81-85

Materials Research Forum LLC

https://doi.org/10.21741/9781644901731-11

Study of the Antarctic Circumpolar Current via the Shallow Water Large Scale Modelling

Kateryna Marynets

Faculty of Mathematics, University of Vienna, Oskar-Morgenstern-Platz 1, 1090 Vienna, Austria

kateryna.marynets@univie.ac.at

Keywords: Geophysical Flow, Nonlinear Boundary-Value Problem, Global Existence, Uniqueness, Positive-Definite Function

Abstract. This paper proposes a modelling of the Antarctic Circumpolar Current (ACC) by means of a two-point boundary value problem. As the major means of exchange of water between the great ocean basins (Atlantic, Pacific and Indian), the ACC plays a highly important role in the global climate. Despite its importance, it remains one of the most poorly understood components of global ocean circulation. We present some recent results on the existence and uniqueness of solutions of a two-point nonlinear boundary value problem that arises in the modeling of the flow of the (ACC) (see discussions in [4-9]).

Introduction

The Antarctic Circumpolar current is one of the five main currents. It is the only current that flows completely around the globe, and is the strongest and largest wind-driven ocean current on the planet. It extends from the bottom of the ocean to the sea surface and is the primary means of inter-basin exchange.

The flow of ACC can be modelled using the following two-point boundary-value problem

$$\begin{cases} u''(t) = a(t)F(u(t)) - b(t), & t \in (0,1), \\ u(0) = u_0, \quad u(t^*) = u^*, \end{cases} \tag{1}$$

where $F : R \to R$ is a given continuous function and $a, b : [0,1) \to [0,1)$ are given bounded continuous functions satisfying

$$\int_0^1 a(s) + b(s)ds < \infty$$

was recently derived as a model for the azimuthal horizontal jet flow components of the Antarctic Circumpolar Current. The existence of nontrivial solutions is of considerable interest, since these correspond to azimuthal flows that feature variations in the meridional direction, being thus models that capture the essential geophysical features, confirmed by field data.

A general result ensures the existence and uniqueness of a solution to Eq. 1, provided that for every $\varepsilon \in (0,1)$ we have:

- all solutions of the initial-value problem

Floating Offshore Energy Devices
Materials Research Proceedings 20 (2022) 81-85

Materials Research Forum LLC
https://doi.org/10.21741/9781644901731-11

$$\begin{cases} u''(t) = a(t)F(u(t)) - b(t), & t \in (0,1), \\ u(0) = u_0, \\ u'(0) = u_1, \end{cases} \qquad (2)$$

exist on $[0, 1+\varepsilon)$ for all $u_0, u_1 \in \mathbb{R}$;

- there do not exist two solutions on $[0, t^*]$ to the two-point boundary-value problem

$$\begin{cases} u''(t) = a(t)F(u(t)) - b(t), & t \in (0,1), \\ u(0) = u_0, \quad u(t^*) = u^*, \end{cases} \qquad (3)$$

for any $t^* \in (1 - \varepsilon, 1 + \varepsilon)$ and $u^* \in \mathbb{R}$.

In particular, global existence for Eq. 2 and uniqueness for Eq. 3 ensure the solvability of Eq. 1. While this is by no means the only possible approach that can accommodate nonlinear functions F, it has some advantages over more classical methods in that the type of hypotheses that need to be verified seem to accommodate large classes of functions.

Modeling of the ACC

For terrestrial regions outside the poles let us introduce the *spherical coordinates* (see Fig.1):

- $\theta \in [0, \pi)$ is the polar angle (with $\theta = 0$ corresponding to the North Pole and with $\theta = \dfrac{\pi}{2}$ along the Equator);
- $\varphi \in [0, 2\pi)$ is the angle of longitude (or azimuthal angle).

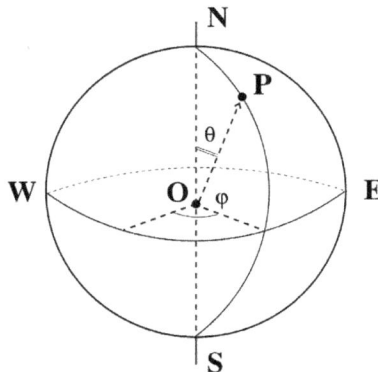

Figure 1: Depiction of the azimuthal and polar spherical coordinates of a point P on the spherical surface of the Earth:

In terms of the stream function $\psi(\theta, \varphi)$, a horizontal ocean flow on the spherical Earth has azimuthal and meridional velocity components given by $(\csc\theta)\psi_\phi$ and $-\psi_\theta$; see [3].

By associating $\Psi(\theta,\varphi)$ with the vorticity of the ocean motion (not accounting for the effects of the Earth's rotation), given by

$$\psi(\theta,\varphi) = -\omega\cos\theta + \Psi(\theta,\varphi), \tag{4}$$

where $\omega > 0$ is the non-dimensional form of the Coriolis parameter, the governing equation for the horizontal flow on the sphere takes the form (see [2])

$$\frac{1}{\sin^2\theta}\Psi_{\varphi\varphi} + \Psi_\theta\cot\theta + \Psi_{\theta\theta} = F(\Psi - \omega\cos\theta), \tag{5}$$

where $F(\Psi - \omega\cos\theta)$ is the oceanic vorticity, while $2\omega\cos\theta$ is the planetary vorticity, generated by the Earth's rotation.

By means of the stereographic projection

$$\xi = re^{i\phi} \quad \text{with} \quad r = \cot\left(\frac{\theta}{2}\right) = \frac{\sin\theta}{1-\cos\theta}, \tag{6}$$

where (r,ϕ) are the polar coordinates in the equatorial plane, we transform the model Eq. 5 in spherical coordinates into an equivalent planar elliptic partial differential equation.

More precisely, using the complex variable ξ, we can write Eq. 5 as

$$\psi_{\xi\bar{\xi}} + 2\omega\frac{1-\xi\bar{\xi}}{(1+\xi\bar{\xi})^3} - \frac{F(\psi)}{(1+\xi\bar{\xi})^2} = 0. \tag{7}$$

Using the Cartesian coordinates (x,y) in the complex ξ-plane, the equation Eq. 7 is equivalent to the following semilinear elliptic partial differential equation

$$\Delta\psi + 8\omega\frac{1-(x^2+y^2)}{(1+x^2+y^2)^3} - \frac{4F(\psi)}{(1+x^2+y^2)^2} = 0, \tag{8}$$

where $\Delta = \partial_x^2 + \partial_y^2$ denotes the Laplace operator; see [3, 4].

Since the ACC presents a considerable uniformity in the azimuthal direction (see the discussions in [1, 4]), we can take advantage of this feature to simplify the problem Eq. 8 further.

Indeed, solutions with no variation in the azimuthal direction correspond to radially symmetric solutions $\psi = \psi(r)$ of the problem Eq. 8.

The change of variables

$$\psi(r) = U(s), \quad s_1 < s < s_2, \tag{9}$$

with $r = e^{-s/2}$ for $0 < s_1 = -2\ln(r_+) < s_2 = -2\ln(r_-)$, for $0 < r_- < r_+ < 1$, can now be used to transform the partial differential equation Eq. 8 to the second-order ordinary differential equation

$$U''(s) - \frac{e^s}{(1+e^s)^2}F(U(s)) + \frac{2\omega e^s(1-e^s)}{(1+e^s)^3} = 0, \ s_1 < s < s_2.$$

(10)

Note that, for $0 < s_1 < s_2$, the change of variables

$$u(t) = U(s) \quad \text{with } t = \frac{s-s_1}{s_2-s_1},$$

(11)

transforms the second-order differential equation Eq. 10 to the equivalent one of the form

$$u'' = a(t)F(u) - b(t), \qquad 0 < t < 1,$$

(12)

where
$$a(t) = \frac{e^t}{(1+e^t)^2}, \quad b(t) = \frac{2\omega e^t(1-e^t)}{(1+e^t)^3}.$$

We will couple the derived differential equation with the Dirichlet conditions on the boundary ∂D of D

$$u(t_1) = u_0, u(t_2) = \alpha,$$

(13)

which reflect the physically relevant condition that D represents the stereographic projection of a surface on the sphere delimited by two streamlines.

Main results
The following result provides sufficient conditions for the global existence of the solutions to the initial-value problem Eq. 2.

Theorem 1. [8] If the continuous function $F : \mathbb{R} \to \mathbb{R}$ satisfies

$$M + \int_0^u F(\xi)d\xi \geq W^{-1}(F^2(u)), \qquad u \in \mathbb{R},$$

(14)

for some constant $M > 0$ and some strictly increasing function $W : [0,\infty) \to [0,\infty)$ with $W(0) = 0$, $W(s) > 0$ for $s > 0$ and satisfying

$$\int_1^\infty \frac{du}{W(u)} = \infty,$$

(15)

and if

$$\lim_{|u| \to \infty} \int_0^u F(\xi)d\xi = \infty,$$

(16)

then all solutions of (2) are global in time.

Materials Research Forum LLC
https://doi.org/10.21741/9781644901731-11

Since Theorem 1 proves the global existence of solutions of the initial-value problem Eq. 2, we now study the question of the uniqueness of solutions to the two-point boundary-value problem Eq. 1.

Theorem 2. [8] If the continuous function $F : \mathsf{R} \to \mathsf{R}$ *is monotone nondecreasing on* R *, then the solution of the Eq. 3 is unique.*

Theorem 3, [8] Assume that the continuous function $F : \mathsf{R} \to \mathsf{R}$ *is monotone nondecreasing and satisfies the* conditions Eq. 14 *and* Eq.16, *for some nondecreasing continuous function* $W : [0, \infty) \to [0, \infty)$ *with* $W(0) = 0$, $W(u) > 0$ *for* $u > 0$, *and subject to the constraint Eq. 15. Then the problem Eq. 1 admits a unique solution.*

Acknowledgements

The support of the WWTF grant MA16-009 is gratefully acknowledged.

References

[1] A. Constantin and R. S. Johnson, An exact, steady, purely azimuthal flow as a model for the Antarctic Circumpolar Current, *J. Phys. Oceanography*, 46 (2016) 3585–3594. https://doi.org/10.1175/JPO-D-16-0121.1

[2] A. Constantin and R. S. Johnson, Large gyres as a shallow-water asymptotic solution of Euler's equation in spherical coordinates, *Proc. Roy. Soc. London A*, 473 (2017), Art. 20170063, 17 pp. https://doi.org/10.1098/rspa.2017.0063

[3] A. Constantin and R. S. Johnson, Steady large-scale ocean flows in spherical coordinates, *Oceanography*, 31 (2018), 42–50. https://doi.org/10.5670/oceanog.2018.308

[4] S. V. Haziot and K. Marynets, Applying the stereographic projection to the modeling of the flow of the Antarctic Circumpolar Current, *Oceanography*, 31 (2018), 68–75. https://doi.org/10.5670/oceanog.2018.311

[5] K. Marynets, On a two-point boundary-value problem in geophysics, *Applicable Analysis*, 98 (2019), 553–560. https://doi.org/10.1080/00036811.2017.1395869

[6] K. Marynets, A nonlinear two-point boundary-value problem in geophysics, *Monatsh Math.*, 188 (2019), 287–295. https://doi.org/10.1007/s00605-017-1127-x

[7] K. Marynets, Two-point boundary-value problem for modeling the jet flow of the Antarctic Circumpolar Current, *Electronic J. Diff. Eq.*, 56 (2018), 12 pp.

[8] Kateryna Marynets, Study of a nonlinear boundary-value problem of geophysical relevance, *Discrete and Continuous Dynamical Systems* (2019), 39(8), 4771–4781. https://doi.org/10.3934/dcds.2019194

[9] Kateryna Marynets, The Antarctic Circumpolar Current as a shallow-water asymptotic solution of Euler's equation in spherical coordinates, *Deep-Sea Research Part II* (2019), (160), 58-62. https://doi.org/10.1016/j.dsr2.2018.11.014

Keyword Index

About the Editors

Ciarán Mc Goldrick is an Associate Professor in the School of Computer Science and Statistics at Trinity College Dublin. He holds a PhD in Engineering from Trinity College Dublin (1998) and has also held Visiting Professor positions at the University of California in Los Angeles (UCLA). He is a Senior Member of both the Institute of Electrical and Electronics Engineers (IEEE) and the Association of Computing Machinery (ACM).

His current research focuses on security and cryptography in both highly scalable and highly constrained devices and settings, and on the timely and responsive communication and control of mobile and actuatable physical devices and settings. He has a strong record in the translation of conceptual research into real systems and realisations, and has been Chief Scientist in a 'high potential' startup in the financial services domain.

He is or has been an advisor to many National and International bodies and agencies, employing both his Research expertise and his strong and enduring commitment to excellence and impact through Learning and Education. He is the Irish representative involved in the definition and formation of future International curricula and standards for graduate education in Networking and Artificial Intelligence/Machine Learning.

Professor Mc Goldrick has co-ordinated, and been in involved in, several EU projects including ICONN, SYSWIND, OCTALIS and ArtWeb. He currently assists the European Commission in various capacities with their proposal assessment and award mandate. He has been awarded as one of "Ireland's Champions of EU Research" by an Irish Government agency for his success and contributions in this space.

Meriel Huggard has been a tenured faculty member in the School of Computer Science and Statistics at Trinity College Dublin (TCD), Ireland since September 2000. She was a visiting Associate Professor in Electrical and Computer Engineering at Bucknell University in 2016. She holds a Ph.D. in Computer Science and a B.A.(Mod) in Theoretical Physics from Trinity College Dublin.

Prof. Huggard's technical research interests incorporate network performance, autonomous and ubiquitous systems. She is also active in the field of engineering education where she is an editor of the IEEE Education Society section of IEEE Access. She is currently an elected member of the IEEE Education Society board of governors.

Her valuable and impactful research has been supported by competitive grant awards from international and national funding bodies, including the European Commission, Science Foundation Ireland and the Irish Higher Education Authority. Prof. Huggard has a broad portfolio of community engagement, including as an external expert reviewer for the European Commission and as a reviewer for a wide range of professional journals and conferences. She is a senior member of both the IEEE and the ACM.

Biswajit Basu is a Professor in the School of Engineering at Trinity College Dublin. He holds a PhD in Engineering from IIT Kanpur (1998) and a Dr. rer. Nat. in Mathematical Physics from University of Vienna (2019). He has also held positions as a Visiting Scholar and Visiting Professor at Rice University USA, a Guest Professor at Aalborg University Denmark, a Senior Marie Curie Fellow at Plaxis BV Netherlands, a Distinguished Guest Professor at Tongji University China and a Distinguished Visiting Professor at Indian Institute of Engineering Science and Technology, Shibpur, India.

He has pioneered the development of time-frequency and wavelet-based algorithms for identification, nonstationary response and control of time-varying and non-linear systems and has published widely in these areas. His current research focuses on nonlinear PDEs, nonlinear hydrodynamics with application to ocean (wind and wave) energy generation and oceanography, and quantum computing with application to machine learning, fluid dynamics, optimization and control.

He is or has been an Editor/Associate Editor or a member of Editorial board of prestigious journals such as Journal of Structural Engineering, American Society of Civil Engineers; IEEE Trans on Sustainable Energy; Journal of Multi-body Dynamics, Institution of Mechanical Engineers UK; Int. J. Structural Control and Health Monitoring.

He has co-ordinated or has been in involved in several EU projects such as NOTES, SYSWIND, UMBRELLA, INDICATE, ICONN and EINSTEIN. He has received several awards of which notable are: President of Ireland EU FP7 Research Champion Award in 2013, Kobori Award for Structural Control in 2014 from the Int. Association of Structural Control and Monitoring and Phil Doak Award from the Institute of Sound & Vibration Research, Southampton in 2015.

www.ingramcontent.com/pod-product-compliance
Lightning Source LLC
Chambersburg PA
CBHW071501210326
41597CB00018B/2649